JOURNAL, DIARY, NOTEBOOK—

ALL NAMES FOR A WAY OF

TRANSLATING THE LIVED MOMENT

INTO A SET OF SIGNS AND SYMBOLS,

MESSAGES TO A FUTURE SELF OR TO

OTHERS WHO MAY PASS THIS WAY.

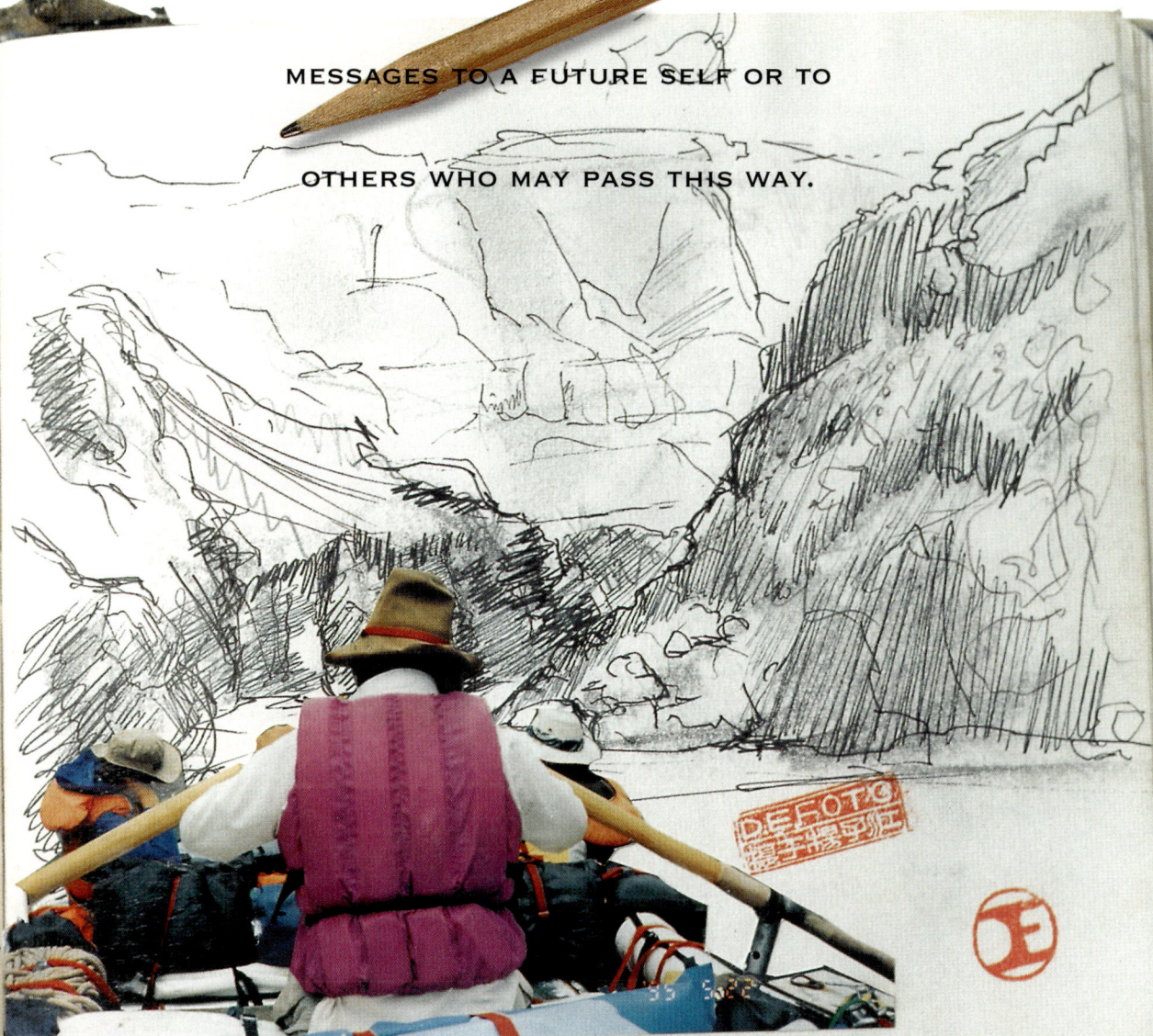

upstream from Deer creek '90

CHEYENNE GIRL 1888

MOTORING 1981 • ROAD RACING

МОНГОЛ ШУУДАН
MONGOLIA
60

ОЛГОСОН
19 ...он ...сар. өдөр
МӨНГӨ ХҮРААГ:
ЧИЙН ГАРЫН ҮСЭГ

/ТЭМДЭГ /

АТГАЛ

НИСЭХ

30   10

САР ӨДӨР   ЦАГ

As the past is enfolded in the present, so we send the present forward to become part of the future.

MONGOL GIRL - ZUUMODT RACE 1992

POLAROID TRANSFER ON DRAWING PAPER AND TOUCHED
UP WITH WATERCOLOR. HÖVSGOL AYMAG. EARLY AM
WHILE VISITING DAVAA AND FAMILY.

NORTHERN MONGOLIA

What can contain our memories — words on a page, a photograph, six smooth stones?

WE GOT A LOT OF MILEAGE FROM THIS CARD

RABBIT WITH
MONGOL SADDLE

MONGOLIA
POSTAGE

Matt's wire
rig for the pots

cabbage   carrots

Ground stove
for tent temper

Chingis snowman
August 14th 1997
upper Goat river camp 2½" snow

ECONOMY CLASS

| FROM | PEK/CHINA | FLIGHT | SEAT |
| TO | ULN/MONGOLIA | OM2232 | 5 A |

Please be present at boarding
gate No 22

MIAT MONGOLIAN AIRLINES

TSAATAN PEOPLE. NORTHERN MONGOLIA. JULY 1993

PIEGAN BLACKFEET. LATE 19TH CENTURY. E. CURTIS.
©DAVID EDWARDS PHOTO

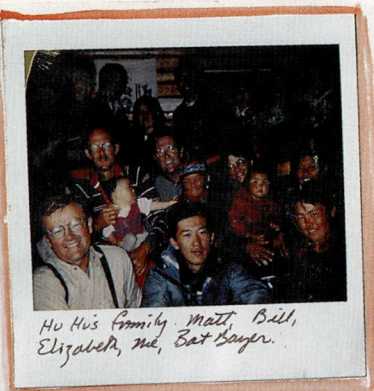

Hu Hu's Family. Matt, Bill,
Elizabeth, me, Bat Bayer.

BE CAREFUL IN THE
TAIGA. TWO MONTHS
AGO A LIEUTENANT
AND TWO SOLDIERS
WERE KILLED
BY BURYAT
POACHERS.
BEWARE OF
STRANGERS.

GOVERNOR. TSAAGAN NUUR.

ЗОРЧИГЧДОД ОЛГОХ ТАСАЛБАР
Дугаар
Уйлчилгээний хураамжинд
200 төгрөг хураав.
199   оны   өдөр
Санхүүгийн тэмдэгтүд
бод хүчингүй

MONGOLIA
80s
МОНГОЛ ШУУДАН

We are creatures who record, who chronicle. We constantly interpret and preserve experience.

The Journal

What makes the journal (or diary, or notebook) so unique is its immediacy. Words strike the page still steaming from the heat of thought. This quality of the eternal present is as recognizable in a hundred-year-old passage from the diary of Emma Sykes that captures the shuffle of customers at Keam's trading post as in the description of place jotted surreptitiously by a solitary coffee drinker in a San Francisco cafe today. This kind of writing has not "had its hands washed for dinner." It retains the rugged sustenance and grit of a campsite supper ten days out on the trail. It is the moveable feast of the moment. Because it is inspired by strong impressions of place and time, and spun from a particular sort of inner solitude, it is a form especially sympathetic to the landscapes and lives of the Colorado Plateau, past and present.

Much recent scientific thinking has to do with systems and the interrelatedness of things. Just as current theory looks less at isolated objects, species, and phenomena and more at overall systems, many contemporary writers seek meaning through impressionistic glimpses that include themselves within the shifting pattern of complex, unique, and unrepeatable moments. In the last spring of the last year of the century, we bring you an issue that celebrates this seeking. It is about place, self, and time.

We are ever grateful to talented collaborators, but for the current issue we thank especially Diane Grua for her assistance in locating historical images to illuminate the printed text; Scott Thybony for his words, ideas, and inspiration; and, especially, the brave and attentive journal-keepers who shared with us powerful pages from their personal chronicles.

CAROL HARALSON

PRECEDING PAGES AND CONTENTS PAGES:

Pages and details of pages from the Mongolian and Colorado Plateau journals of Dave Edwards and a stack of his journals. All photos of journals and journal pages, objects, and travel memorabilia by Tony Marinella.

ontents

WINTER 1998 - 1999

SCOTT THYBONY

Certain places draw out the journal-keeping impulse more than others, and the Colorado Plateau is one such place. Writer Scott Thybony explores this impulse through "place notes" from a photographer, a Navajo sojourner, and his own compactly jotted observations.

The heavy slap and amphibian smell of water, the cliffs rising an oar's-breadth away, the sense of eternality within constant change — the river has impelled many to the written word. Here are slices of that experience in the writings of seven travelers, each with a different reason for entering the river, each with a distinctive voice for telling the tale.

JOINT PUBLISHERS OF PLATEAU JOURNAL

Museum of Northern Arizona

Grand Canyon Association

PLATEAU PARTNERS

Arches National Park

Arizona Strip Interpretive Association

Bryce Canyon Natural History Association

Canyonlands National Park

Canyonlands Natural History Association

Capitol Reef National Park

Capitol Reef Natural History Association

Colorado National Monument Association

Dinosaur Nature Association

Dixie Interpretive Association

Dixie, Manti-LaSal & Fishlake National Forests

Entrada Institute/Friends of Capitol Reef

Glen Canyon Natural History Association

Grand Canyon National Park

Hubbell Trading Post National Historic Site

Kaibab National Forest

Mesa Verde Museum Association

Mesa Verde National Park

Museum of Western Colorado

Northern Arizona University, Cline Library

Peaks, Plateaus and Canyons Association

Petrified Forest Museum Association

Petrified Forest National Park

Public Lands Interpretive Association

USDI Bureau of Land Management

Wupatki, Sunset Crater Volcano and Walnut Canyon National Monuments

Zion Natural History Association

PLATEAU JOURNAL
PUBLISHED BY GRAND CANYON
ASSOCIATION AND THE MUSEUM OF
NORTHERN ARIZONA

VOL. 2, NO. 2, WINTER 1998-99

PLATEAU JOURNAL

Editorial and Design
Carol Haralson, Designer and Editor
L. Greer Price, Editor

Pamela Frazier, Partnership Coordinator
Michele Madril, Museum of Northern Arizona
Tracey Hobson, Mesa Verde Museum Association
Paula Hosking, Petrified Forest National Park

Specialized Assistance
Faith Marcovecchio, Grand Canyon Association
Kim Buchheit, Grand Canyon Association
Vince Grout, Marketing and Development

Photo Archives and Library
Diane Grua, Cline Library, Northern Arizona University
Sara Stebbins, Grand Canyon National Park
Tony Marinella, Museum of Northern Arizona
Kim Besom, Grand Canyon National Park

. . . and warm thanks to Scott Thybony for his special assistance
in the development of this issue of *Plateau Journal*.

This publication is made possible in part by support from
Grand Canyon Railway and the **National Park Foundation**

35 IN THEIR OWN
WORDS

JUDITH FREEMAN

The experiences of Lucy Flake, Rachel
Lee, and Emma Sykes, recorded in their
diaries one hundred years ago, are as
compelling today as when they were
written. In them these three remarkable
women give us a glimpse of
Mormondom of the 19th century, of
pioneer life, and of how the frontier
seemed to a forty-year-old newlywed
from England.

*I write on my lap with the wind
rocking the wagon.— Algeline Ashley*

from *Women's Diaries of the Westward
Journey*, by Lillian Schlissel

PRODUCED BY THE MUSEUM OF NORTHERN ARIZONA AND THE GRAND
CANYON ASSOCIATION, *Plateau Journal* is a semiannual publication. Subscription
to *Plateau Journal* is a benefit of membership at designated levels in the Museum of
Northern Arizona and in Grand Canyon Association, and in some Plateau Partner
organizations.

SUBSCRIPTIONS AND GIFT SUBSCRIPTIONS are available independent of
membership. Contact the Museum of Northern Arizona, 520-774-5211, extension
222.

TO PURCHASE COPIES please contact Grand Canyon Association, 800-858-2808.
Single copies are available for $9.95 and special discounts are available for quantity
purchases. Back issues are available.

PLEASE ADDRESS CORRESPONDENCE TO *Plateau Journal*, Museum of Northern
Arizona, 3101 N. Fort Valley Road, Flagstaff, AZ 86001. Telephone 520-774-
5211, extension 216.

Contents

LFORD XP2

I got to explore schist and
water and ocotillo to
touch it and think about
how it looks and feels.
I got to step on rubble
washes and look at all
the different rocks, feel
horsetail and hold cactus
spines. I looked at my
hands holding these
beautiful things and my
whole world was richer
because of it. You made
me feel like another
beautiful part of that
already beautiful place.

THE NOTEBOOK

Our talk drained rather quickly off into silence and we lay thinking, analyzing, remembering, in the human and artist's sense praying, chiefly over matters of the present and of that immediate past which was a part of the present; and each of these matters had in that time the extreme clearness, and edge, and honor, which I shall now try to give you; until at length we too fell asleep.

JAMES AGEE, FROM LET US NOW PRAISE FAMOUS MEN

L*ate in the day* I find myself staring at the desk, trying to remember. The grain of the wood swirls back and forth, moving in eddies and crosscurrents the way a river flows. I'm trying to remember the Colorado River at flood stage. It's a different river chugging muddy red than when it runs clear. But the detail has slipped away. Whenever I have questions about the river, I end up talking with my friend Dave Edwards, boatman and photographer.

Inside his three-tiered studio, contact prints hang next to postcards and sketches. A cowboy hat sits on a Mongolian saddle above camera gear and stacks of books. An assortment of rugs and weavings from various nomadic groups adds splashes of color. Despite the apparent randomness, all the pieces fit. Squeezing past bags of clothing collected for the orphans of Ulaan Baatur, I find Dave packing for a photo shoot in Mongolia.

He sets aside his work to answer my question about muddy waves. "They hit harder," he says. "They have more punch." He begins to give me examples and stops. Running downstairs, he digs through a river bag and hands me a blue notebook. The cover is worn with use and water-stained; the binding repaired with duct tape. "Here," he tells me, "take this. It's all in there."

Certain places draw out the journal-keeping impulse more than others. From the first European encounter with the Colorado Plateau, travelers have found it compelling enough to write about. A Spanish soldier, Pedro de Castañeda, chronicled the Coronado expedition in 1540. Others have followed his example. They recorded the distances covered, routes taken, waterholes found — what you might expect of outsiders passing through unfamiliar country. But they also noted something more. Their writing reveals a struggle to come to terms with the vast spaces of a new world and the strange, the wide-eyed other they found themselves facing.

Explorers and missionaries brought back journals; Mormon pioneers and an occasional trapper kept diaries. Women wrote with insight about the

Certain places draw out the journal-keeping impulse more than others.

ABOVE: A Grand Canyon party, summer of 1890. Bean-Tappan Collection, Cline Library, Northern Arizona University, NAU.PH.660.121.

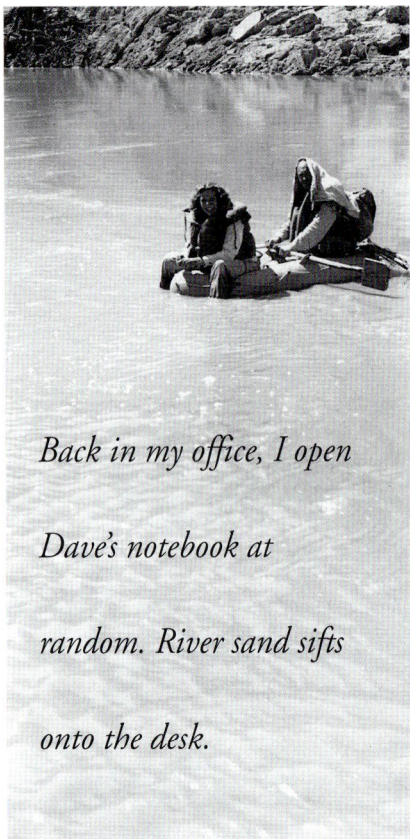

Back in my office, I open Dave's notebook at random. River sand sifts onto the desk.

human tragedies and daily sacrifices of life on the trail, often glossed over in accounts left by the men. And in the wake of John Wesley Powell, everyone floating the Colorado, it seems, took notes.

Back in my office, I open Dave's notebook at random. River sand sifts onto the desk. "Security is mostly a superstition," I read. "It does not exist in nature, nor do the children of men as a whole experience it. Avoiding danger is no safer in the long run than outright exposure. Life is either a daring adventure or nothing —Helen Keller." I remember the quote. It's one Dave read at a memorial service for Dugald Bremner, a friend who drowned in a river accident. Returning to the front, I begin leafing through the notebook.

It's part scrapbook, part naturalist's journal, part reference. Like his studio, it draws from whatever sources make sense. Pages overflow with observations ranging from river hydraulics to using meat tenderizer for easing a bee sting. He has filled others with American Indian poetry, notes on nautiloid fossils, a line drawing of a dazed passenger "right off the bus," and a recipe for red chili sauce. A diagram of useful knots lies tucked in loose alongside the sketch of a tangled rope. He has taped a dragonfly to one page, as flat as a bug on a windshield, and a feather on another. A list reminds him of the points to cover in his passenger orientation talk. He's gathered quotes from astronauts and another from Einstein on the mysterious nature of the universe.

"There are no rules about keeping a notebook," Dave told me, "except one. You have to share it with friends." Continuing to read, I find several pages filled with raven lore. He gives an accounting of items stolen by these birds – soap bars, sandwiches, a halibut steak – with the dates and incidents duly recorded. And river details, ones I've forgotten or only half remembered, cover page after page. He describes the feel of a boat as it breaks into the current, the hungry rise of the river, the joy of good company. Each line reanimates old memories.

This type of notebook has a rawness to it, a spontaneity. Nothing is lashed to a storyline. This is especially true of Dave's travel journals from Mongolia. They burst with local color, incidents, characters. Watercolors, photos, beer labels, and postage stamps crowd the same page. He sketches street scenes in the capital city, documenting the breakdown of social order – black marketeers, orphans, pickpockets, corrupt officials. When he finally escapes to the empty steppe, it's a release of spirit. A portrait of an eagle hunter fills one page, his face weather-hardened. Other watercolors show children racing Mongol ponies and reindeer nomads swallowed up by endless expanses.

THIS RIVER RUNNING BRINGS OUT STORIES

FROM THE JOURNALS OF DAVE EDWARDS

Sept 1990. There is a different feel about a river on the rise. It's nothing noticeable at first. Then bits of driftwood and after that long strands of dirty white foam
Finally, it takes on a lumpy look, swelling, sloshing as though gaining appetite.

The light tapping of rain drops on a tent fly.
Silver tape holding my ripped tent together.
Knarled, wrinkled Russian pennants hang from the dome.
From the Chuya rally in South Central Siberia.

notes 1990
fragrance of tamarisk. limestone like coral
ground coffee
water drips on hot rocks . . .
washing in cold river water
cotton clothes that dry fast in a hot wind . . .
a letter from my former lover that says no words, only
lipstick lip prints. It gets wet in my pocket . . .
violet green swallows in a light rain.
smell of muddy water. slippery clay.

my fingers hurt. Every cut & nick has been stretched open in the dryness. My hair is matted and dirty and my body is salty.

Winter trip 1991. February.
Pale light. Thin clouds. Light seems to have no source.
Frozen sand crunches like snow. 2 ravens hang out
on my boat. The variety of sounds they make amazes me.
Snores, chortles, croaks, clicks, clippy guttural imprecations.
Dirt on beaks from burying food.

~ the boat goes to the top of the eddy and you
turn into the current pulling as if to go upstream,
the current seizes the boat and whips it around,
water sloshes over the metal deck and the boat shudders.
I like that.

~ I have never known any experience where there is so much laughter at breakfast, lunch & supper day after day after day. This river running brings out stories. They rise like freed chickens, flopping, squawking, sending us all howling. It is good to be weak from laughter. *'94*

Horse Tail   Nactura

River Stones   Yucca
Found Objects

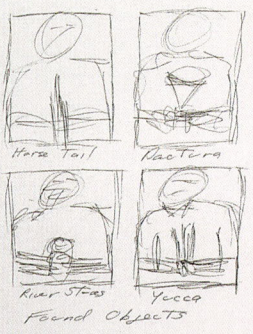

Try to do a series of 4 or 5 photos of Susan holding various objects in her hands. Try to use the same pose the same look — with some variation

Later,
I did some photos for this series — but I'm not sure I was successful in producing a series — I made some photos that are individually good — but I'm not sure they work in pairs — let alone fours — and I used the last of my sepia film

Here's what's wrong — All of the photos need to be of a similar scale and composition

"If you always succeed you are not trying enough."
Woody Allen

I hoped you liked the pictures you took at Nankoweap. I liked the objects you were choosing for me to hold, but it was I'm thinking about what those looked liked and how to hold them. By this time I think I realized how much you love humanity and that maybe you take pictures just to make people look really beautiful.

Your pictures of rocks and plants are truly wonderful, but I loved a lot within. You appreciate hands, arms, shoulders, and faces along with rocks, plants, and places. I was glad I could contribute. I really liked the picture of you and me. I hope you had great take.
You look truly great in it.

Cobble bars — I wonder at them — marvel at them — these rocks have been worn by thousands and millions of years — how long does it take to become a grain of sand — how long does it take jagged rock falls to become collections of rounded rock — how far have they traveled — where did they come from? where are they going —?

I got to stop of rubble washes and look at all the different rocks

Susan Mile 200 on the right

The skeleton of a plant shows the tracery of life here in the bottom of the canyon

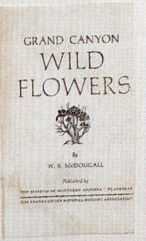

Plants bloom in spring — up to summer and leave behind brittle remains

GRAND CANYON
WILD
FLOWERS
by
W. B. McDougall

OCCOTILLO
I like the shape of the word — and the way it sounds! OCCOTILLO (Fouquieria splendens) "Spiny shrubs which drop their leaves as soon as the soil dries but quickly produce a new crop when the rains come. Very showy when in bloom!!" —

Showy indeed. They have scarlet blossoms — atop thorny stalks that resemble the barbed wire of a maximum security prison compound more than a plant. OCCOTILLO

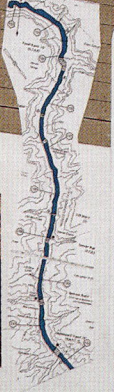

This picture I liked best. From that ground it was the eye level my shoulder above the slope of the rock and I'm only part of the subject.

I looked at my hands... holding these beautiful things and my world was all the richer for it.

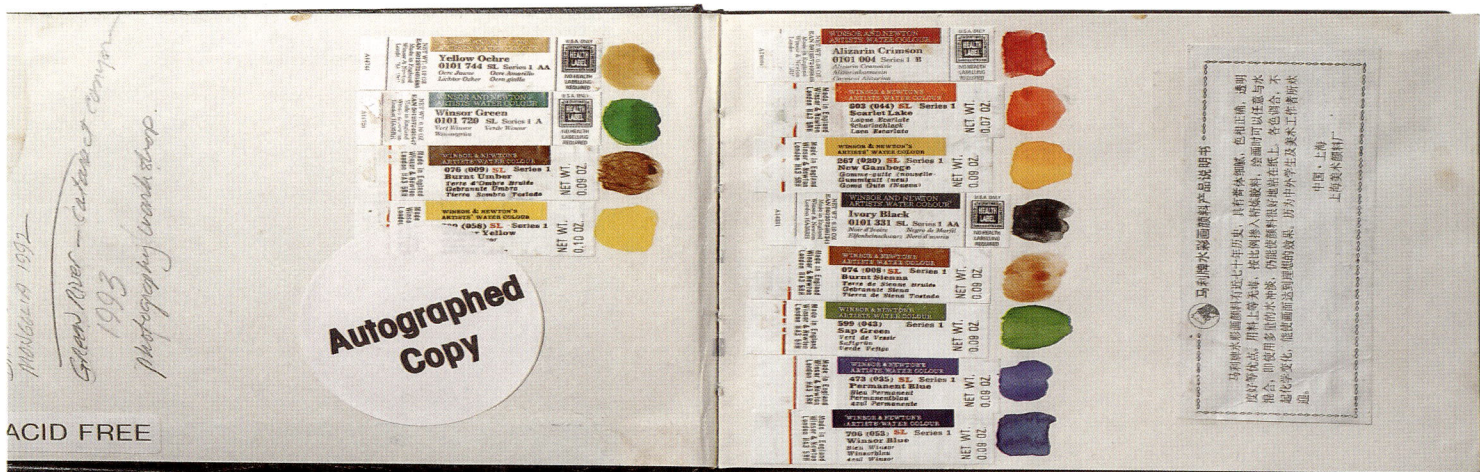

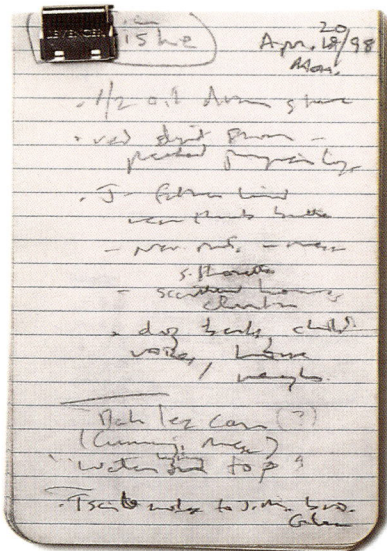

When I ran the Colorado River, the lead boatman who trained me couldn't see the point of keeping a journal. We pulled into camp one evening, and he spotted another guide writing in a notebook. "He must think what we're doing is important." But what you write about in a journal doesn't have to be world shaking. Its force comes from what we take for granted. Insignificant in itself, the ordinary connects to a wider pattern and seen in the right light transforms the everyday into something personally meaningful. The small truths, the ironies of daily life, the humor – these turn up in a good journal. You recognize one when it's done well, like hearing a hollow thump when you find a ripe melon in a pile of green ones.

After discovering Dave's notebooks, I began asking other friends if they kept journals. A surprising number did. One of them, Peg Swift, touched ground on the Navajo reservation for a number of years, working as a writer and photographer. Her journal has a simple beauty and wide-open rhythm, matching the high desert where she lived. When we talked about journals, she remembered once picking up a hitchhiker on his way west. He told her about the loneliness of always being on the highway and how keeping a journal helped him. "When things are bad," he said, "you can read back and see they've been worse."

The small truths, the ironies of daily life, the humor – these turn up in a good journal. You recognize one when it's done well, like hearing a hollow thump when you find a ripe melon in a pile of green ones.

Geologist John Maxson across from Crystal Creek, Grand Canyon, 1937. Photo by Robert P. Sharp. Carnegie-Caltech Collection, Cline Library, Northern Arizona University, NAU.PH.94.27.83.

TOP: Shonto: Scenes in floor of valley, 1932. NAU.PH.413.662.

RIGHT: Navajo cornfield, south of Inscription House, 1927. NAU.PH.413.75.

FACING PAGE: Monument Valley, 1927. NAU.PH.413.126.

Photos by Philip Johnston. Philip Johnston Collection, Cline Library, Northern Arizona University. Johnston conceived the idea of Navajo code talkers during World War II.

"*Tsech'ishi,* say it slowly. *Tse* – rock; *ch'izhi* – rough.

Sixteen miles from the place of Many Farms. Squash and pumpkin in the fields. Irrigation ditches and dry blowing soil, and sometimes it doesn't rain enough, but plants grow; they grow beautifully.

Here, at Tsech'izhi, nothing is cultivated. Sheep and a few goats graze at the foot of the mountain – sage, black grama, four wing. The sheep eat, grow fat, grow wool, grow lambs for the ceremonies – year after year.

In fall the piñon seeds ripen. The people spread blankets under the trees, shake the trees, and the loosened nuts tumble to the ground. Money for the trader; food for the winter.

+

When it rains the arroyos fill and this house becomes an island. I stand in the rain, let down my hair, wash my hair in the rain. Watch a tumbleweed caught in water rush by, and rocks and stones, and red earth move from here to there across this land, to small streams, to larger streams, to the Colorado River, the Sea of Cortez.

This rain clarifies, focuses my world, and confines me to this one place, for a time.

After the rain: the dust settled, the land clean, clear, sparkling in the setting sun. Mid-summer, the smell of sage strong. Sage mingling with piñon smoke from the Yellowhair's hogan by the windmill and the bells of sheep and goats coming home from the mountain.

The sky becomes blue again.

+

I look all around. Above me, blue sky; below me, redrock canyon. Sheer sandstone cliffs reach down to the canyon floor. Hogans, cornstalks in winter fields, bare cottonwoods. A dog drinks from the semi-frozen wash. Smoke rises from a hogan, a thin blue curl in the early morning sky; and above the smoke, a hawk, circling in old wind currents, watches for that one rabbit too slow.

In winter I cook for the big ceremonies: the Night Chant, the Mountain Top Way. Cut up mutton and potatoes, onions and chile. Mix flour, baking powder, salt, and water in a big tub, cover with a bluebird flour sack, and let it rise. Heat lard in a smoking pan. Boil large pots of coffee.

The buttes of Monument Valley are red gold, only for a second, at dusk. The plain beyond is shimmering, and then dark beneath the setting sun.

One by one the stars appear. I ride up the mountain in the back of Kee's pick-up, his two children nestling in my lap. Tonight is the last night of the Fire Dance. The first killing frost is past. A ring of pale light around the moon.

I sit inside the circle of spruce boughs; watch the white-clayed clowns, brandishing fire-sticks, make jokes about the people standing around laughing. A central fire of cedar and piñon, the flames reaching for the sky. Thirty dance teams from across the reservation: the feather dance, the fire dance, the corral dance. Fifty sheep slaughtered and 500 people gathered: turquoise, silver, the rich colors of Pendleton blankets; new pick-ups and wooden wagons; the smell of wood smoke, roasting mutton, frybread, hot coffee. The prayers and dancing continue until dawn.

+

I ride the black horse, from deep well tank to the mountain and back, racing across the plain, no fence to stop us. When thirsty we drink cool mountain water from the wash. When tired we rest in the shade of an old cedar. Bluebirds fly around us, laughing, teasing, telling us to go faster, faster still. This horse is of the wind.

We ride around slowly; look out across the land, look in all directions, for the cattle, the sheep, small dots in the shimmering distance. Cattle graze lazily in the hot afternoon sun. Sheep rest in the shade of canyon walls, wait for the coolness of evening.

+

At night we lie on sheepskins on the earth floor and watch the fire embers slowly die, and watch the stars cross the summer sky through the smoke hole of the hogan. Great white shooting stars, the messengers of this hot August night. In the morning we awaken to bird song, the earth floor cool, dew on the grasses and wildflowers, and the world lightening in the east.

:   Giving rancher Pete Espil a lift back to his house, I ask how Deadman Flat got its name. "Well," he says, "I heard it got named for a drunk cowboy who couldn't catch his horse. Every time he got close to it, the horse spooked and ran off. He kept chasing it for a couple of hours. The longer he chased it, the madder he got. Finally he drew his pistol and shot it. He felt so bad about what he'd done, he turned the gun on himself and pulled the trigger." — DEADMAN FLAT, ARIZONA

:   "They're coming now!" shouts a young Hopi boy in a camouflage T-shirt. Far below the mesa, flute dancers shift into single file and start up the trail. We cross the mesa to the old pueblo where the dancers will end their journey. Next to a truck with a bumper sticker reading, "Everything Is Sacred," the boy picks up a tattered toy rabbit and holds it to his ear. It's playing a plinky, metallic rendition of the "Battle Hymn of the Republic."

"It's singing!" the boy says.

"What song?" I ask him.

"It's the flute dancers' song." — WALPI, ARIZONA

:   On a drive across the reservation, I tell anthropologist John Farella about some graffiti on the wall of a gas station at the last stop. "Ghost Dance," it read. "The ancestors will live again."

We talk about the movement that swept through many of the western tribes a century before and the current interest in collecting Ghost Dance artifacts. "The same people who laugh at Ghost Dance shirts stopping bullets," Farella says, "will pay $250,000 for one."— FIVE-MILE WASH, ARIZONA

:   Mule riders sit in the shade of a piñon, taking a water break at Cedar Ridge. A woman asks the head wrangler if he's married. "Yes," he tells her, "for three years. My wife's well educated but has to work below her abilities up here. She's a fine woman but she lived a spoiled life before I found her. I made her send her credit cards back to her father. But each time I threaten to send her home, he sends us more money."

Satisfied with his answer, the woman turns to the other wrangler. "Tell us about your life," she says. He sits there, head bent without speaking, pushing a pebble through the dust. The woman repeats her question. He takes a deep breath, "What chapter do you want?"— GRAND CANYON, ARIZONA

:   An old Navajo, dressed in jeans and a flannel shirt, pumps gas at the Marble Canyon trading post. Clouds are building above the Kaibab Plateau to the west. I ask him if it's going to rain. He squints at the sky and shrugs.

"I don't know," he says. "I didn't watch TV this morning." — MARBLE CANYON, ARIZONA

NOTEBOOK, JOURNAL, DIARY. . . I shift from one term to the other in a single conversation. Formal differences may exist, but in practice they become interchangeable. I think of diaries and journals as following a rough chronology, with a diary tending to be more private than the others. A notebook is whatever you bring to it.

People keep journals for different reasons. Some try to make sense of the world by recording the incidentals and waiting for a pattern to emerge. A reflective act, journal keeping lets you trace the currents of your own life. Sometimes it functions as a safety valve, sometimes as a way to come to terms with suffering and death. Scientists in the field keep them, monks in their cells, artists living on the margins. Journals can be intensely private or outrageously public. Some people keep their work under lock and key, and others post it on the Internet. Either way, you write for an audience, even if the reader is only yourself, years later.

When I worked on the river, the closest I came to a journal was an envelope of loose notes. Months later, I'd enter some of them in a notebook. Only after beginning to work as a writer did I keep what resembles a journal. The need emerged after realizing most of the stories I told friends when returning from assignments never made it into the article. They were too off-the-wall or didn't fit the thrust of the story. So I began taking my rough notes — terse, cryptic scrawls — and transcribing them in a narrative form I call field notes. These notes allow me to circulate material before publication to catch mistakes and, more importantly, to let ideas flow. They need to tumble and clack together, the way boulders move in a flash flood. If momentum stops, they settle to the bottom. And finally silt up.

Keeping notes trained me to see events, even ordinary ones, in terms of a story. From these, I extract what fits the assignment. The rest sits there waiting to flesh out the ghost of memory, the way journals do.

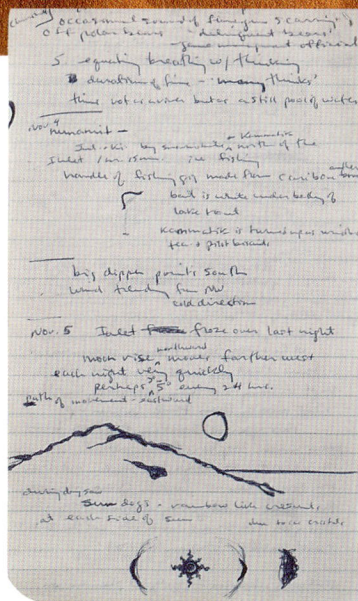

JOURNAL OF SCOTT THYBONY

Five lupines growing in a straight line atop a sand dune, Monument Valley Tribal Park, Arizona. Photo by Fred Hirschmann.

FACING PAGE: Navajo reservation. Photo courtesy of Peg Smith.

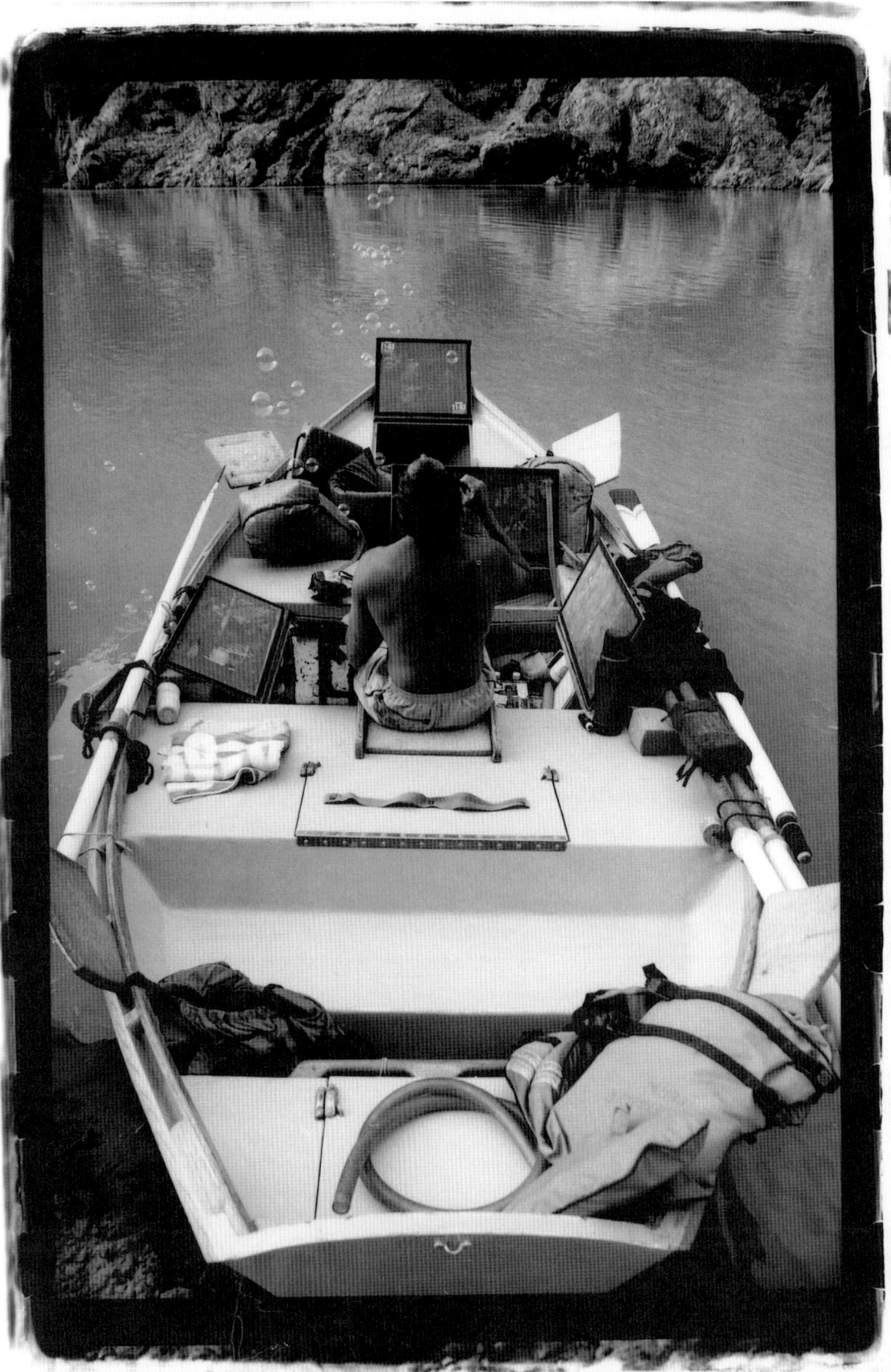

PHOTO BY RAECHEL RUNNING

REMEMBERED RIVERS

Some people are drawn to a blank page. Entering a bookstore, they gravitate to the shelf with journals containing empty pages, nothing more. They find themselves attracted by the idea of filling space with words and drawings, by the thought of making their own book for their own reasons. River guides tend to keep journals, or wish they had kept one. The Colorado makes its way onto the page as often as it gets in the blood. Many notebooks carried down the river have a strong visual element, ranging from pictographic sketches to the full kaleidoscopic effect. These illustrated notebooks have a vibrancy, a visual impact not found in blocks of text alone. A printed book, with its orderly lines and solid regularities, follows conventions meant to be taken for granted. The surprises come in the writing style, the unexpected turns of plot, or the compelling ideas presented.

But with a visual notebook, you don't know what will turn up next. Different forms blend. Words wrap around sketches and found objects land in the middle of watercolors. Each page becomes a slice in an open-ended composition. You might find a pen and ink drawing with a photo glued on top, a scrap of letter taped next to a colorful food wrapper, wildflowers pressed between pages, fragments of maps, or an expired identity card. You can learn a lot about people by what holds their interest, by the things they drag home to the bone pile, by what ends up in their journals.

The first book I bought with my own money was *The Notebooks of Leonardo da Vinci.* The great Renaissance artist liked to think with his pen, rushing to get an idea on paper without concern for linear sequence. Drawings and words rub shoulders, filling page after page. Calling his notebooks "a collection without order," Leonardo let his thinking run its natural course, unchecked. You see this in his life-long interest in the nature of water. He studied all aspects of it, entranced by the percussive force of falling water or the pattern of converging streams. The artist noticed how currents twist together like plaited hair, like the tendrils of a plant. In one of his early notebooks, he sketched a river flowing down a mountainside, and in his last ones drew swirling, apocalyptic floods.

Water continues to hold a fascination for those who spend their lives close to it. In a boatman's journal you might find a sketch of a breaking wave or a note on hydraulics, echoing Leonardo's observations 500 years earlier. Rivers and journals, both have a reflective quality and a restless energy.

Water continues to hold a fascination for those who spend their lives close to it. In a boatman's journal you might find a sketch of a breaking wave or a note on hydraulics, echoing Leonardo's observations 500 years earlier. Rivers and journals, both have a reflective quality and a restless energy.

Rod Sanderson 93.17A ✓

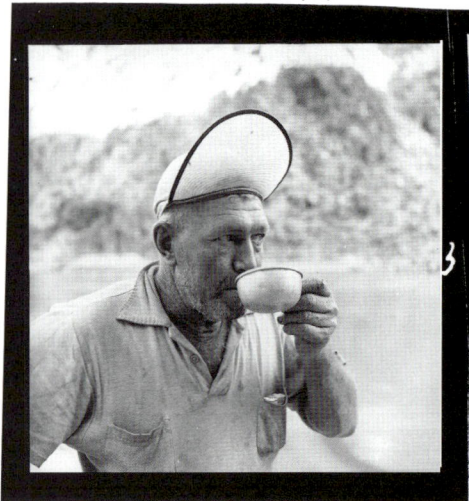

✓ 93.25 B Frank Masland

Jorgen Visbak ✓ 93.13B

✓ Dock Marston 93.37A

✓ Josiah Eisaman 93.53A

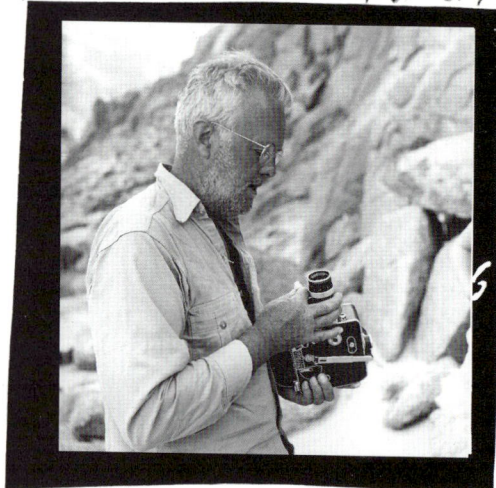

The following excerpts come from visual notebooks kept on the Colorado River. They follow the loose sequence of a river trip, beginning at the put-in and ending at the take-out. Many of the notebooks are on-going works, difficult to borrow during the river season when actively in use. For most of these journal keepers, time on the water is a return to a place they have never quite left. For others, it's wild and new, still a river of the imagination.

Following are selections from the river journals of a handful of seasoned contemporary boatmen, a scientific adventurer from the 1930s, and one recent first-timer.

The final entry comes from the most famous river journal, written by John Wesley Powell, who kept a daily record of his epic first descent through the Grand Canyon in 1869. Twenty years later he reworked these spare notes into a journal with illustrations, published as *The Exploration of the Colorado River and its Canyons*.

"It is highly probable that the snapshots we make of our lives and times today will become vital historical documents of the future" wrote riverman, teacher, and photographer Bill Belknap. While perhaps not vital, these five images, captured in 1954, are a valuable part of the tradition of river portraiture. In photographing our guides and fellow passengers, we document the grittiness of our hair, clothes, and coffee, the lack of formality, and the manner in which camping reduces life to its essence.

CLOCKWISE FROM TOP LEFT: outboard motor pioneer and outfitter Rod Sanderson; businessman, explorer, and author Frank "Fisheyes" Masland; river historian and boatman Otis "Dock" Marston; passenger Josiah Eisaman; and chemical engineer Jorgen Visbak. Photos by Bill Belknap. Belknap Collection, Cline Library, Northern Arizona University, NAU.PH.96.4.93.17A, 93.25B, 93.37A, 93.62C, and 93.13B.

Spanning more than a century and all levels of experience, these accounts share a deep connection with the Grand Canyon and the river cutting through it. They converge in the telling, twining together the way water flows.

Major Powell's story of descending the Colorado River bears only a rough resemblance to his original diary. For example, the book devotes more than five pages to the most dramatic incident of the trip when three of his men abandoned the expedition at Separation Canyon. In his diary for August 28 he simply records, "Boys left us. Ran rapid." But those few words became the catalyst for his classic tale written years later. A journal holds words the way duff holds a spark. An idea flashes, gone before you can think it through. But put it in a notebook and it smolders, waiting to flare up with the slightest breeze.

Mottled sunlight on wall of Marble Canyon, upstream from Nankoweap, Grand Canyon National Park, Arizona. Photo by Fred Hirschmann.

DAVE EDWARDS, a
Colorado River guide,
has lost count of the
trips he's made. More
than 150, he thinks. At
some point, each
separate trip becomes
part of a larger journey.

On his website
(www.daveedwards.com)
Dave posts photographs
of the places covered in
his travel journals.

beach talk. Lee's Ferry.
introduce guides. talk. talk. talk. talk.
life jackets, your friend, rainwear, drinks for
the day, urinate in H2O
basic safety. drink water.
sand in boats. no smoking on boats or in
kitchen.
demonstrate highsiding, bailing, line, where
not to sit.
+
Dumb Questions.
Which way is downstream?
Did the government put all this sand down
here or did it get here some other way?

Is the dam at the end of the river or the end
of the lake?
Are we on this side of the river or that side?
+
Canyon walls etched black against the sky.
The moon coming up piercing your eyes like
a flashlight in the face.
— night sounds. water. crickets. wind. smell
of rain.
+
The feel of the moving boat, the lift, plunge,
shudder, twist, bump & even rip.
+
I like the water muddy, red desert brown. I
like the look of the rapids. The roar is more
resolute. Earth & water.

View from Nankoweap of Saddle Mountain, Grand Canyon
National Park, Arizona. Photo by Fred Hirschmann.

JACK STARK

Sights we'd like again to see,
     Frank's arms pointed in a V.
As upon the forward hatch he stands and looks the rapids o'er,
     And to hear his warning shout
     As he turns his boat about
To dodge the rocks and feel the waves wash on the decks once more.

JACK STARK, a geologist at Northwestern University, kept "The Doggerel Log of a Canyon Trip" on a geology expedition sponsored by the Carnegie Institution of Washington and California Institute of Technology in 1937.

The whole story's never told,
We'll continue till we're old
Remembering new details and perhaps we'll add a few:
    But its no exaggeration
    Our sincere appreciation
Of friendliness and comradship of river men we knew.

And on many a wintry night,
By our home fires burning bright,
With a grandchild on our knee and our slippers on the rug,
We will think of now and then
Wishing we could hear again
The Canyon's echoing cry of: "Its a plug! plug! plug!"

L'ENVOI

When we cross the River Styx,
With Old Sharon at the helm,
Knowing not what distant port of call we'll end on,
    May we find, as once before
    Welcome waiting on the shore
From the friendly Little Gray Men of the Canyon.

One of these, a Scot, named Ian,
Called by all his friends, the Peon,
A professor in the California Institute of Tech,
He made little fires to sit by,
But when no one else was near by,
Lighted huge bon-fires of driftwood, higher than a
                                      giraffe's neck.

Johnny Maxson was a savant,
World wide traveler and bon vivant
Famous for his blufferitis and his pseudo-fossil finds.
  And in wilds of Nankoweap
  Up and down its slopes so steep,
His time was spent in tracking Norman Ethan Allen Hinds.

Jack possessed an absent mind,
Like professors of his kind,
Leaving gloves and hammers on Archean rocks
  Frank supplied gloves that were swell
  But Jack missed his pick like hell
Till he found it where Bob hid it in his bedroll with his socks.

Next the baby of the crew,
Robert Sharp of Harvard U.,
Climbed so fiercely that his shoes lost every nob.
  He would leap from crag to crag
  "Firm not brutal" like a stag
Till he won deservedly the title of Daredevil Bob.

ABOVE: The Doggerel Log captures, colorfully and playfully, the human aspects of a trip with a distinct scientific purpose: to systematically study the igneous and metamorphic rocks that compose the inner gorges of the Grand Canyon. Stark prepared the original log during the river trip, then created hand-drawn copies for each member of the expedition and for Buzz Holmstrom, who passed the group on his daring solo run down the Colorado River.

The Cline Library at Northern Arizona University houses two Doggerel Logs, those that were given to Holmstrom and to geologist Robert P. Sharp. The one reproduced here was Sharp's and is part of Carnegie-Caltech Collection, MS 293. Reprinted by permission of Sarah T. Klare.

The pages shown above chronicle the four geologists who made the entire trip. Maxson was a colleague of Campbell's at Caltech; Sharp, a Ph.D. candidate at Harvard University.

Jack Stark "the Rykrysp King," Frank Dodge, and Ian Campbell on the *Fairchild,* 1937. Campbell, a Caltech geologist, was the organizer of the expedition. Dodge, who had rowed on and/or participated in four trips in Grand Canyon, was one of the most experienced boatmen available. It was "aboard the forward hatch of the good ship *Fairchild*" that Stark composed his log. Photo by Robert P. Sharp. Carnegie-Caltech Collection, Cline Library, Northern Arizona University, NAU.PH.94.27.117.

KATE THOMPSON

It all started on my 1st dory trip . . . Being close to the water, feeling the solace of a sleek craft, weighing the balance of such a boat, I felt at home. We glide through the riffles, the dory is a sounding board as the oars clunk against her gunnels. I love boats. All kinds of boats. But mostly really small ones with nice lines.

+

The lateral that I had to burst through surfed me out into the current & down the throat of gigantic waves. The crescending water seemingly gave way for us, but by the 4th wave, it was fearsomely huge, & I knew we could flip end-over-end. We make it through & start rejoicing (in our minds) & then we were struck with the realization that the 5th wave was too huge to even comprehend. We stare in awe at the beauty & magnitude of the universe converging into one, complete wave. We were going to flip & I plant the oars deep in one last weak attempt to get us through. The whole boat crashes through the peak of the wave & we stall . . . at the very crest of the wave.

I remember seeing such incredible beauty, so in awe of the power. Such indescribable beauty as our small boat stalls out at the crest, ready to sink backwards into the wave which we just ran.

+

The Canyon walls are on fire. Such passion they emit, beckoning to me. I yearn to be part of this beauty, the quality & refinement of light.

KATE THOMPSON, a geomorphologist who has run rivers throughout the West, recounts her first trip in a wooden dory and a run through Hermit Rapid.

JOURNAL OF KATE THOMPSON

OCOTILLO. I like the shape of the word – and the way it sounds. OCOTILLO.

+

Ocotillo regenerates itself after each rain, putting forth vivid green leaves on its thorny stalks soon to be followed by brilliant scarlet flowers.

+

Mile 98.5. Each time I stop to scout Crystal rapid, I stare at it. Commit it to memory. Rehearse the run in my mind. Immediately after running the rapid, I think, "A B C, Alive Below Crystal." But then, as Wesley Smith cautions, we are always, "above crystal."

+

Cobble bars – I wonder at them – these rocks have been worn by thousands and millions of years – how long does it take to become a grain of sand – how long does it take jagged rockfalls to become collections of rounder rock – how far had they traveled – where did they come from – where are they going?

+

Here in the Canyon's depths silhouettes often reveal more than direct illumination.

JOHN RUNNING has floated the Colorado more than fifty times, keeping a photographer's notebook. He's also taking the journal-keeping tradition in a new direction, if direction makes any sense in cyberspace.
His "Colorado River Log" website (www.thecanyon.com/johnrunning) combines photographs and text in an electronic form.

RAECHEL RUNNING (www.thecanyon.com/rmrunning), a photographer like her father John, began river running at the age of ten. Dave Edwards taught Raechel a few basics on her first trip rowing a boat down the Colorado. "Get it right," he told her, "get it right, keep it straight, get it right." She follows these words in life as well as on the river. Raechel signs on with a river crew whenever she can break away from the studio.

JOURNAL OF JOHN RUNNING (DETAIL); PHOTO OF JOHN RUNNING BY RAECHEL RUNNING

RAECHEL RUNNING

Winter is a red river running fast past naked trees. Cool air rises & blows; ice forms & the metal is cold. There's an art to winter dressing & it seems we all are living in our many layers. The ritual of dressing & undressing, donning warm & dry gear, eliminating wet socks, wet booties, pants, & gloves. The black bags take on new forms and new odors as well. . . At night we sit around the fire drying our gear. The gloves, socks, & boots dangle from well positioned sticks over the fire drying. After the meal is prepared we eat quickly to ensure the heat of the meal remains so; the warm nourishment downs itself through our hungry mouths and the fire crackles, & the drying goods steam.

We are making 35 miles in a day. Moving approx. 4.5 miles an hour. A steady pace. Quick glances & reminiscing over the hikes we've made, the storms that have passed, the camps we've called home, and the years of our lives we have shared in this place of timeless beauty & wonder.

+

The moon rose gracefully, a luminous orb in a slow & ritualistic dance. A mist of transluminous cloud drapes the round shoulders of the moon & rising in an arc of concentrated monuments, the dark canyon walls served as a frame holding this canvas of sky. The river began to sparkle, the light penetrated the dark shadows of a still night & lit up our laughter. To be surrounded by such beauty, & in the company of good folks, in song, in laughter our lives lay woven by the currents of the river.

+

Yesterday when we got into camp a beautiful storm passed through the canyon. Everything down river became a mist of gray blue silhouettes & when the rain finally arrived the sun still shone. Because of the light, the rain looked like snow falling white. We all dove for raingear & hunkered for shelter.

+

It's amazing to think of the reality of all we know & have created could very well become dust in the wind. That in less than 500 million years everything could be eroded from the surface of the earth as we know it. No skyscrapers, highways, houses . . . no trace of being here. Hmmm.

+

I am sure that people take the Grand Canyon experience with them. It's funny to me how some of the people I've met & encountered still come to mind. Their eyes & sunweathered faces. The brief exchange, a catalyst for years of thought provoking ideas, ringing in the memory along with the sound of slow water gurgling in the company of crickets.

+

Glassy waves, silver threads of nearly invisible current lines, soft murmurs of geology in the making, the smooth V of channeled water which rushes into the crashing confusion of exploding waves . . . Focus, hold the oars firm and really put your body into it with a prayer upon your lips as you struggle to keep to the line of froth . . . Gold light drips from the rhythm of the dip, pull, feather twist of the worn oar in callused hands. . .

+

We are now approaching Lava. Just now we can hear the roar of water. Miraculously the sun is out, shining on us . . . may the sun shine on us forever. Its warmth and its light; may we go through this "window" in beauty.

Day 3 Will this be the day I jettison my watch?

+

Day 7 They hit a wave in an unnamed rapid that tossed the chub up on its side and threw him off. He surfaced & looked for the boat – the first rule when you hit the water is to find the boat (and he couldn't find it in the waves). He got sucked thru a wave. "It's so dark in a wave," he told me. "You have no idea how dark." He surfaced again, still couldn't find the chub. He got sucked thru a second wave: then he made the chub & was drug to safety. "It's the first time I've been afraid of the River," he said. That's the difference between us tourists & the guides, I think – that they know enough to be afraid.

+

Day 9 This was the day for Lava Falls – which we reached about 4:30. The flooding had changed it, and this, the wildest rapid on the river, was more brutal than ever. . . Mad Dog got knocked off his line and dove into a hole as big as – well, as big as the Grand Canyon. When he knew the boat was going to flip, he dove into the waves. I remember seeing the boat spit up out of the hole like a squeezed watermelon seed. It stood on its nose and all I could see was its shiny black bottom.

+

Day 12 We packed our dry sacks for the last time this morning. For twelve days, everything we have needed or wanted has fit into these grocery-bag-sized black rubber containers. When we unpack them into our car in the parking lot at Pierce Ferry, we will be returning to a more material life, filled with things and diffusion.

When we left camp, we made a flotilla out of the rafts to float the hour to Separation Canyon. Hal sang "The Rivers of Texas," Keith read to us out of his book, Frances sang "500 Miles." The mood was somber but very sweet.

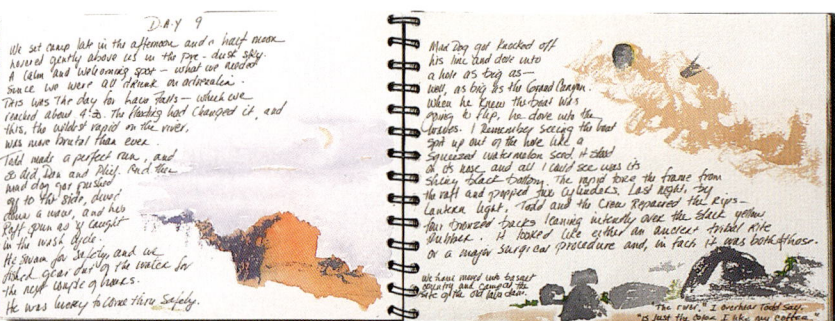

TERESA JORDAN, author of *Riding the White Horse Home*, recorded her first river trip down the Colorado as part of a group of women writers whose impressions of the experience are gathered in Northland Press's recent anthology *Writing Down the River*.

JOHN WESLEY POWELL

August 29. We start very early this morning. The river still continues swift, but we have no serious difficulty, and at twelve o'clock emerge from the Grand Canyon of the Colorado.

+

Ever before us has been an unknown danger, heavier than immediate peril. Every waking hour passed in the Grand Canyon has been one of toil. We have watched with deep solicitude the steady disappearance of our scant supply of rations, and from time to time have seen the river snatch a portion of the little left, while we were a-hungered. And danger and toil were endured in those gloomy depths, where ofttimes clouds hid the sky by day and but a narrow zone of stars could be seen at night. Only during the few hours of deep sleep, consequent on hard labor, has the roar of waters been hushed. Now the danger is over, now the toil has ceased, now the gloom has disappeared, now the firmament is bounded only by the horizon, and what a vast expanse of constellations can be seen!

The river rolls by us in silent majesty; the quiet of the camp is sweet; our joy is almost ecstasy.

....

JOHN WESLEY POWELL is credited with making the first journey through the Grand Canyon on the Colorado River in 1869. The journey lasted nearly three months, and only six of the original ten men emerged at the mouth of the canyon.

WRITING A NOTEBOOK lets the hand train the ear. Write enough and you begin to hear the scratches, the distortions that remind you to go back and try again. Another entry, another attempt. Each day you come back to a blank page, and each day it appears as solid as a desktop without a nick in the surface. Then you notice the grain moving in ripples and waves, and let it carry the writing forward. You begin to write by dividing each thought into slices one word thick. You string them together, page by page, until the storyline emerges.

My own notebook contains observations arranged in a loose narrative. Over time the recorded incidents grow less important than the pattern, the lines of force holding them together. The act of writing shapes how I see; the pattern becomes visible in the telling. The notebook lets you see each encounter in terms of a story. Even the growth of a leaf follows a narrative pattern from the opening bud to its wind-carried close.

Everybody you meet has a story, but only some find the storyline. What begins as a few troubled thoughts, entered in a notebook late at night, leads you slowly outside yourself into something larger. You have a sense of closing the distance. The telling itself carries you beyond the narrow present, the way a boat breaks out of the eddy and a stronger current takes hold.

Colorado River reflections between Nautiloid Canyon and 36 Mile Rapid, Grand Canyon National Park, Arizona. Photo by Fred Hirschmann.

Sunday at 2 oclock Meeting was held Singing, prayer by H. Barney Bro. Pheem
E. H. Groves. & A. G. Ingram Spoke from the Stand on the test that was near
at hand alluding to the troops &c. a good spirit prevailed Ben by E. H. Groves

13th   This Morning a greate number of Indians returned from an
expidition South west. also with Bro J. D. Lee. Meeting was
convened in the afternoon,

20th   This Bishop and J. D. Lee went to Conference at S. L.
City. in the afternoon Meeting was held only a few of
Saints convened Many being absent. Henry Barney Presiding
on the 16th an Indian child was blessed before it expired
by the Bishop its name was Matilda.

27th   Met at 10½, Sing prayer by W. Littlefield, E. H. Groves. Benj. Platt
A G Ingram addressed the Meeting Benediction by J. R. Davis
In the afternoon the brethren & Sisters testified to the truth &c.
H. Barney first Recouncillor Spoke down on tithing. Ben. by A. G. Ingram

Thursday
Oct 1st   This being fast Day was kept by the brethren and Sisters turning out
to Meeting generally A good feeling prevailed.

Sunday
        ____ the Meeting by Singing prayer by E. H. Groves, after Singing
        ____ in Council Viz Saml White and Saml Leigh

In Their Own Words

THE DIARIES OF LUCY FLAKE, RACHEL LEE, AND EMMA SYKES

In the winter of 1877 a small party of colonists in six covered wagons left their comfortable homes in the fertile river valleys of Utah Territory for an unknown land. Leading the party was an able frontiersman named William Jordan Flake, and with him were his two wives Lucy and Prudence. ARIZONA OR BUST was scrawled on one of the wagon covers.

The going was tortuous, over roads that were little more than crudely blazed trails. Often the party halted while the men made the route passable with picks and shovels. Forage was often scarce for the herd of over 200 cattle and fifty loose horses, so frequent stopovers were necessary.

When the party reached the Kaibab Plateau in northern Arizona, Lucy Flake looked out across the sandy, barren plain stretching before her and saw a solitary wagon floating like a white speck against the brown desert. As her party approached the wagon, they could see a man pacing back and forth beside it. His wife sat in the wagon box cradling a dead child. In her diary, Lucy Flake describes what happened next:

> We camped nearby and I went over to see what I could do. The poor mother was helpless in her grief. The little body had to be washed and prepared for burial. I had never touched a dead person before except my own precious babies whose graves I had left behind. I shrank from the task but someone had to do it. William took a board from one wagon and one from another wagon box until he got enough to make a little box to bury the child in.

This act of kindness nearly cost the lives of Lucy's older daughters, Mary and Jane, who fell sick a few days later from fever and a constriction of the throat. Lucy used all the remedies she knew. The girls gargled with vinegar, salt, and pepper, and Lucy bound pieces of fat meat to their throats. Yet it seemed both would perish. As Lucy prayed in the wagon for her daughters, a group of travelers overtook the Flake party. Among them was an aged blind healer, "Aunt Abbie" Thayne.

The journals of three extraordinary women who settled on the Colorado Plateau add another dimension to our view of the region's pioneer history. Their stories, each very different, give us an idea of the price they paid for their independence.

Page from the diary of Rachel Lee. The Huntington Library, Art Collections, and Botanical Gardens, San Marino, CA

Aunt Abbie was escorted to the wagon where Jane and Mary lay. No one had described their ailments to her, "but as soon as the blind woman put her head inside of the wagon cover she sniffed the air and said, 'Diphtheria.' then she went to work. In her wagon she had medical herbs for every disease and in her head had the knowledge of their use and our darling girls were saved through her ministrations. One of the things she did was to put poultices of grated carrots on their lungs. . . I had never been around diphtheria before and had not known that the little body I had prepared for burial had died of that disease."

The diary of Lucy Hannah White Flake is one of many remarkable accounts by 19th-century female settlers on the Colorado Plateau. The women wrote of their lives and relationships, their medicinal knowledge and sisterly bonds, and the experiences that tested the strength of their constitutions and resolve. Reading their words, we see the beauty and the terrors of the landscape through the lens of a feminine eye.

Like women everywhere, in every age, the female pioneers who homesteaded Utah, Arizona, and Colorado were foremost concerned with the welfare and survival of their families, and their diaries reflect this preoccupation. Familial relationships, whether with husbands and children, or even with polygamous sister-wives, take center stage in these journals. But they also afford an extraordinary glimpse of the native people encountered by newcomers, and of the social and cultural life of the fledgling pioneer societies.

From the beginning, Mormon settlers from Utah had found the region to the south harsh and inhospitable. A scouting party of seasoned explorers sent to Arizona by Brigham Young in 1873 had met with failure. The country was too dry and too hot, they reported. It would not support settlements. Anxious to extend the reach of Mormondom, Young tried again a few years later: this time he decreed that families, not men alone, were to head south. Women meant community, meant true settlement of the land. Where men had failed, women and children might succeed. Lucy describes the journey vividly.

The Road

W omen were the actuaries of the road, tallying the miles with the lives that were lost.

—LILLIAN SCHLISSEL

Spring dust storm enveloping a Utah juniper, floor of Monument Valley, Arizona. Photo by Fred Hirschmann.

FACING: Old farming equipment at Lees Ferry. Photo by Paul Berkowitz.

Utah, west of St. George. Photo by Bill Belknap. Belknap Collection, Cline Library, Northern Arizona University. NAU.PH.96.4.218.4.

Lucy Flake

Winter came early. Indians whom we met said it was the coldest for years. I shall never forget the 14th day of December, 1877. We had only made a mile and a half that day. The wind and sleet were terrific. 'there is no use. the teams will not face this storm. Find as sheltered a place as you can and pull in.' William's voice rang out above the wind as he delivered this message to each of the teamsters. He was riding ahead, trying to point the loose stock [who] were on the verge of stampeding. . . As soon as the poor shivering animals were unharnessed, they huddled together in what shelter the six wagons made. The men and boys who could be spared from the herd climbed into the wagons with the women and children. . . The wind blew so strong all day that it seemed the wagons would be overturned. . . Heavy snow fell just before we reached the Little Colorado River. It was fortunate for us that this extreme cold did not strike us on the desert. We would have frozen to death as wood was so scarce. . . At Black Falls we found another family whose child, an eleven-year-old boy, had died. Again, I prepared the body for burial. The water was so cold that it froze on the body as I washed it.

Lucy Flake was born in Illinois in 1842 and came west with the great migration of Mormons, walking from Missouri to Salt Lake City. Her family eventually settled in Beaver in 1853, and there, at the age of fifteen, she met William Flake, six years her elder. His family had made the brutal trek across the Mojave Desert in 1851 to become part of the Mormon colony that settled in San Bernadino. By the time William was fifteen, both his parents were dead and he had already established himself as a stockman with a fine herd. When the so-called "Utah War" erupted in 1857 and the California Mormons were called back to Utah, William returned to Beaver. Soon after, he and Lucy were married.

Lucy's journal shows us a young woman setting up house with a new husband for whom she clearly has passionate feelings:

Our first home was two little log rooms, with very little in them besides love. The bedstead was built in the wall, had a rawhide strap laced back and forth for 'springs'. . . it was so high from the floor that Will had to pick me up and put me on it. I could manage to scramble off by myself. We cooked on the fireplace. We served ourselves from the frying pan or bake skillet. We had two pewter bowls, a butcher knife, a big spoon that he whittled out of wood, a teaspoon, three two-tined forks and a silver ladle. We had a good shuck tick and wool mattress on our bed, plenty of bedding. That and a stool or two to sit on was about the amount of our housekeeping outfit, but we had LOVE, other things could be acquired.

That love would soon be tested, first by the deaths of two babies, though four others born to the couple would survive. An even harder test was to come when, one evening after supper, William drew up his chair toward Lucy.

> Taking my face in his hands he turned it around so that he could look into my eyes and asked, 'Lucy, dear, could you share your husband with another woman?' I thought at once that he was joking, so laughingly answered, 'Sure, if I could still retain first place in your affections.' He bent his head over until his lips met mine. Each kiss carried the same thrill the first one had. He stood up and pulled me to him, and I noticed a seriousness about him that I had never seen before as he said, 'Lucy, I have been counseled to take another wife, if you are willing.' I could not speak, nor could I keep the tears out of my eyes. . . 'Don't try to answer me now,' he said in his gentlest voice. . . 'Think it over, pray over it, as I have and then let me know.' I flung my arms around his neck, and held him, as though I would never let him go. . . Of course I was not willing. He was mine. Mine by all the laws of man and God. For ten years we had been all the world to each other. We were made for each other. Why should I let someone else come between us? Neither of us spoke again, but I could tell by the tears that dropped on my bowed head that he was suffering as much as I was.

When William returned from a trip a few days later, Lucy had decided. "Will," she said in greeting to her husband, "who is the young lady we are going to marry?"

The "young lady" turned out to be Prudence Kartchner, a sixteen-year-old beauty whom Lucy had watched grow from a child. In a passage marked by extraordinary empathy, Lucy wrote, "It must have been rather hard for a good looking, fun loving, popular young lady to marry a man eleven years older than herself, who already had a wife and was the father of six children, and yet any girl should have felt proud and honored to become the wife of William Jordan Flake."

When Lucy, William, and Prudence arrived in Arizona, they stopped at a settlement near Joseph City called Allen's Camp. The camp settlers were living the United Order—all their resources were pooled for the common good. Meals were taken communally at one long table, and the women took turns cooking, waiting table and dishwashing. But dissension broke out within the camp over communal property, and William Flake decided that his family should establish a place of its own. In June he set out by horseback to seek a possible place.

While William was away, Lucy set up house in a wagon box on the ground. The colonists were trying to build a dam on the Little Colorado, and the weather was foul.

> I had long realized how futile it was to try and build a dam in that treacherous stream of shifting sand. We could not raise anything and were eating up what we had brought with us. The spring winds filled the air with that fine sand until one couldn't see two rods away. . . I was so homesick and blue, I found that the tears were very near the surface but I was seldom alone long enough to indulge in a good cry.

Lillian Schlissel, in *Women's Diaries of the Westward Journey* (Schocken Book,s, 1992), describes the wagons and provisions that carried mid-19th century travelers across rugged country.

The Wagon

Building a wagon and provisioning for the trip were major undertakings. The overland wagon had to be built of seasoned hardwood to withstand extremes of temperature; an ordinary farm wagon was not strong enough to stand up to . . . hard traveling. . . . The wagon and animals might cost four hundred dollars, the largest single expense of the expedition. The wagon was built to be amphibious. A tar bucket hung from the side of each wagon, and the slats were caulked for river crossings. The covering of the wagons was a double thickness of canvas as rainproof as oiled linen, or muslin, or sailcloth could be made to be. Wagon tongues, spokes, axles, and wheels were liable to break, and most emigrants traveled with spare parts slung under the wagon beds. Grease buckets, water barrels and heavy rope were essential equipment. As wagons deteriorated from overloading or breakage, repairs were made.

Rachel Lee

The Flakes' youngest son George had always been a frail child, though Lucy describes him as "exceptional. . . with a mentality of a child twice his age." George fell ill. For weeks Lucy watched him suffer until she felt so overcome she prayed that God might end his ordeal. When her husband finally returned from his trip, he learned that little George had died a few hours earlier. The child was buried in a crude coffin which Lucy painted bright blue.

Into the heaviness of this grief, William delivered his news: he had found a new home for the family, a ranch on Silver Creek owned by James Stinson. Stinson was willing to sell, but the price seemed unattainable. Eventually, through a series of inventive deals whereby future crops and herds were promised to Mr. Stinson, the Flakes bought the ranch, and they set off for their new home. When Lucy first saw it, she was overcome.

As far as the eye could see, the rolling hills were covered with waving grass. One large house and a line of small adobes nestled in a little brown patch among tasseled corn and ripening bearded barley in the center of the valley below. A few cottonwood trees and willows fringed the banks of the silvery stream that gave it its name. . . . It seemed this beautiful valley was a bit of Heaven reserved for us as a reward for all we had suffered. . .

The Flakes prospered, acquiring more land which they sold to other families, and sharing their good fortune freely with newcomers. The town they established was called Snowflake, after a church elder named Erastus Snow, and William Jordan Flake.

NOT ALL WOMEN who came to the Colorado Plateau had such happy endings to their stories. One of the most tragic tales centers on the figure of John D. Lee and his wives, a number of whom kept journals.

Lee is perhaps remembered best for two things—first, for his involvement in the bloody affair known as the Mountain Meadows Massacre, and second, for establishing the first ferry crossing of the Colorado River at the site which still bears his name.

Many facts surrounding the Mountain Meadows Massacre are now known, though for years the incident was shrouded in lies and half-truths. In 1857, a wagon train from Arkansas and Missouri was attacked by Mormon colonists together with a band of Paiutes, and some 130 men, women, and children were

The Paria River flowing into the Colorado River, past Lonely Dell, 1921. Lees Ferry is upstream on the Colorado River—just out of view on the left side of the photograph. Emery Kolb Collection, Cline Library, Northern Arizona University, NAU.PH.568.1383.

Crossing the Colorado River at Lees Ferry. Emery Kolb Collection, Cline Library, Northern Arizona University, NAU.PH.568.6491.

killed just northeast of present-day St. George. Very young children were spared and taken into the homes of Mormon settlers. Reasons for the incident remain the subject of conjecture. That Lee was a participant, however, is without question. And, although it is unlikely that he was its instigator, he alone was tried for it.

Before he was brought to trial, however, Lee eluded federal marshals for almost twenty years. During this time he established the ferry at a remote spot on the Colorado River. There he lived with two of his wives, Emma and Rachel, in the place they named Lonely Dell.

Rachel Andora Woolsey Lee was Lee's sixth wife. Rachel, two of her sisters, and her widowed mother all had married Lee in the days before the great exodus to the West. In all, Lee had nineteen wives—an astonishing figure even in those free-wheeling polygamist times—though many of his wives became dissatisfied and ended up leaving him over the years. Rachel remained faithful to the end, even arranging to live with Lee in his prison cell in exchange for doing the back-breaking work of cooking and serving as laundress for the other prisoners.

In 1853, four years before the Mountain Meadows Massacre, Lee led a party of Mormons south from Salt Lake City to establish a settlement called Fort Harmony, between present-day Kanarraville and New Harmony. Rachel Lee was twenty-eight years old at the time and had two children (two others had died in infancy). Her diary is largely a record of meetings held at the fort—both religious gatherings, at which the colonists were exhorted again and again to live righteously, and community sessions where practical matters were discussed. Her original diaries, like those of her husband (some of the most vivid and detailed journals of the time) are now part of the collection of the Huntington Library in San Marino, California.

To a modern eye, Rachel's journal is remarkable for what it excludes. She records that her husband left on a trip to Salt Lake on September 20, a week or so after the events at Mountain Meadows (Lee was surely headed there to deliver a report on the affair to Brigham Young), but records no trace of the occurrence or its effects on the community. The only hint is the entry for September 13, which reads, "this morning a greate number of Indians returnd from an expedition South west. also with Bro J.D. Lee. . . " That "expidition" was likely to the meadows.

Perhaps Rachel had no knowledge of what happened at the meadows, but what of the orphans who suddenly turned up for adoption? And what of the booty that came into the community—the dozens of shoes, dresses, and other items of clothing, the farm implements, and hundreds of head of cattle?

On October 17, she writes: "this Evning the Bishop and J.D. Lee returnd from Conference [in Salt Lake City] Bringing with them Some news that the Soldiers ware on the way &c they say thare is a distemper raging North called the 'Horse Distemper'. . . J.D. Lee was afflicted with it." One is struck by the stiff tone and the reference to Lee by his surname, but most by what Rachel does not record: that Lee returned from the trip in the company of Emma, his newest wife, whom he had met, courted, and married in the two weeks he had been away.

Rachel's journals are more forthcoming about community life, as in this entry for October 1857:

> In the afternoon peter Shirts made confession of whiping Mary Morse. Also, Mary Morse made confession of using abusive Language twards P. Shirts. both ware forgiven after some explanations. . . alma Lee and Sml E Groves confesed thier faults in playing and cuting up in the way they did one Eveing in the carrel in runing about in a state of nudity &c Some good counsel was given to the youth.

As revealing, and of even greater interest historically, are Rachel's entries recording the baptisms of Indian children who were purchased by the Fort Harmony colonists—the Paiute, or less often Ute, boys and girls who often had been traded for little more than a jar of beads and half a side of beef.

The practice of selling Indian children for slaves had been going on in the Southwest long before the Mormons arrived. For years companies of Mexicans had made regular trips north to trade for children, often stolen by the Utes from weaker bands and later sold to traders for use in the Mexican mines. The Mormons disapproved of the practice but when their opposition resulted in tensions with tribal leaders, Brigham Young decreed that colonists should buy as many children as possible to keep them from a worse fate.

A passage from the preamble to an act passed in 1852 to offer relief to Indian slaves and prisoners (*Laws of Utah,* 1852) described the plight of these children: ". . . they are carried from place to place packed upon horses or mules [and] lariated out to subsist upon grass roots or starve, and are frequently bound by thongs made of rawhide until their hands and feet become swollen, mutilated, inflamed with pain and wounded; and when with suffering, cold, hunger, and abuse, they fall sick, so as to become troublesome, are frequently slain by their masters to get rid of them. . ."

Diaries, records, and family histories have, from the beginning, been important to Mormon practice. In the January 1, 1867, issue of the *Juvenile Instructor,* published in "Great Salt Lake City" for young Mormons, an elder exhorted the young to preserve their times:

Keep a Journal

Wishing to impart something for the benefit of our young friends, the readers of the Juvenile Instructor, I will commence by conversing a little with them upon the subject of keeping an account or journal of events as they daily take place around them. . . . Let all the boys and girls get them a little book, and write in it almost every day.

But the object is not so much to get you to keep a journal while you are young, as it is to get you to continue it after you become men and women, even through your whole lives . . . for you live in as important a generation as the children of men ever saw. . . .

If men had not kept a journal in former days, we should not now have any Bible, Book of Mormon, Doctrine and Covenants or any other book. . . . And it is so with the history of all nations. There has been a record kept of the wars, troubles and difficulties [we] have had to pass through. Afterwards, men take these journals and accounts and make a history of them. . . .

In a diary entry for April 27, 1856, Rachel wrote: "some children (Indian) ware blessed being purchased by the brethern Moroni Ingram." A note adds that one of the children, barely a year old, died five days later.

Before Lee's capture and trial, Rachel had homesteaded for several years at a place called Jacob's Pools—later renamed Rachel's Pools—a day's ride east from Lees Ferry. She was often alone for months at a time with only her children for company, dealing with Indians, caring for a herd of cattle, and trying to eke a living from the harsh land. Later, she joined Lee in "Moweabbi" (present-day Moenave, near Tuba City), at the farm he established there while hiding from authorities. Lee's journals from the time describe the visits he and Rachel made to Chief Tuba and his wife at Moenkopi, where they shared the delicacy of a fresh melon and made small talk.

Rachel's diary ends in 1860, but we know from other accounts that she faced many trying times in the seven years between the end of this diary and the death of her husband by execution in 1877.

It's easy for us now to pity polygamist wives like Rachel Lee. But there's this to consider as well: they were often extremely strong women who led very independent lives for at least a part of the time, engaging in business dealings and controlling property left in their care while their husbands traveled between homes. One cannot help but think of Rachel Lee as a woman of exceptional resources and courage, capable of setting out alone on horseback or driving a balky ox team, traveling across a country that had frightened the hearts of many robust men. At Rachel's Pools today many of the low rock walls she built to fence off the springs still stand as a testament to her labors.

AN ENTIRELY DIFFERENT PICTURE is afforded by Emma Walmisley Sykes, who left her lush, green native England for life in the Sonoran desert.

Emma came by train to Holbrook, Arizona, as a new bride in July of 1895. She was the eldest daughter of a comfortable, middle-class Victorian family, and her diary reflects a more worldly sensibility than that of Lucy or Rachel. Yet one senses there was nothing pampered about her; nor was she timid about embracing the rigors of her new life. The day after arriving in Holbrook, she set out by horse and buggy with her husband, Godfrey, for Keams Canyon Trading Post where the couple would spend the next seven months. For the first six weeks, Emma faithfully kept a "honeymoon diary" in which she recorded impressions of her new home.

Godfrey Sykes was an Englishman who as a young man embarked on a life of wandering and fell in love with the American West. Eventually he settled near Flagstaff where he ran a machine shop and a small cattle operation with his younger brother Stanley. Godfrey met Emma during one of his return trips to England and, after a prolonged correspondence, the pair decided to marry. Godfrey accepted the invitation of his friend and fellow Englishman Thomas Keam to spend his honeymoon at Keam's trading post while Keam made an extended journey overseas.

Emma was forty years old when she came to Arizona; her husband was six years younger. Nothing had prepared her for the landscape she encountered when she arrived. On the first day, en route to Keams Canyon, she wrote that she "never could have imagined anything so awe-inspiring as this. . . magnificent scenery all around and beautiful blue above."

Emma Sykes

Emma Sykes brings the fresh perspective of a true outsider, one equipped with a keen eye and considerable descriptive skills. Her diaries were penciled in pocket-sized penny notebooks purchased from a London stationer. On her first night camping under a bright Southwestern moon, she lay awake listening to the terrifying sounds of horses munching grass nearby, but she and Godfrey arose refreshed. "This I attribute to having taken so much pure air. I have come to the conclusion [that] at home we hardly know what real pure air is like."

The next day the newlyweds stopped at the Indian store at Bidahochi, where Emma recorded her first encounter with natives: "They appear to be very peaceful and gentle and are talking softly in such a soft musical language, which reminds me of water gently dripping or the riffling of a rivulet. One came close to me and smiled and looked with keen interest into my bag and I showed them my tan shoes. . . I am really enjoying all this so much and am so happy."

The "tan shoes" are only a part of the wardrobe and accessories Emma toted out West. She describes heating up her curling iron on the stove, wearing a fine blue dress which pleased Godfrey, and a riding habit that included a fetching hat. Contrasted with the meager inventory of the new bride Lucy Flake (a single wooden spoon her husband had carved, and three forks), Emma would know few deprivations in her temporary new home. The impression is heightened by the diary entry describing her arrival at the trading post:

Mr. Keam met us at the gate. . . and I was very much struck with the hearty greeting. He is exceedingly nice. . . Well, after washing and dressing I put on my blue cotton frock. We had a delicious breakfast, very tender little ribs of mutton, potatoes and onions done together, rice pudding, bread and butter made here, very nice tea, milk and various sauces and beans peculiar to the country. Then they showed me the house. It is like an Indian bungalow. The sitting room walls are a blaze of color, it is just like a small museum—plaques, potteries, baskets, rugs, blankets, beads, photos, dolls, feathers, and all sorts of curios. . . the rooms all open one into the other. Although Mr. Keam does not leave until Wednesday he is moving out of his big room for us at once . . .

Emma Walmisley Sykes on the porch at Keams Canyon, 1895; Keams Canyon, 1895. Photo by Adam Clark Vroman. Photos courtesy of the Sykes family.

Aspen on the Kaibab Plateau, Arizona. Photo by Fred Hirschmann.

Over the following months Emma Sykes recorded in vivid detail the stay at the post. A cook and servants left her free to roam the nearby canyons with Godfrey, who undertook the job of teaching her how to ride horseback. The store was a two-minute walk away, up a sandy wash—"a straight shed-like building with windows and doors, a long counter and a lower counter full of sand where the Navajo and Moquis [Hopi] stand and smoke as they do their trading. They are very slow, walk in—stand about—seem to be deep in thought and then just walk away, perhaps never doing any trading."

It must have been very hot in July, though initially Emma finds little to complain about. The impression she gives is of one fascinated by the people and landscape around her:

> We have had a little stroll up the canyon; it is a most wonderful place, truly grand. Imagine yourself surrounded completely by rocks and small cedars. When you look up they seem to almost touch the sky, and the perfect stillness is so remarkable. . . . Just now a party of women and a baby crying [have] just ridden up on their burros . . . they move slowly, and are very graceful in some of their movements, especially when they motion with their hands and fingers in pointing to places at a distance.

Within a few weeks Emma began to more squarely face the realities of life at the trading post, including the realization she would be "Alone a great deal," as she wrote wistfully. "Godfrey busy at the shed, felt very lonely, but must be sensible and feel sorry I gave way to such feelings."

Emma's loneliness was eventually relieved by friendships with the archaeologist Dr. Jesse Fewkes and his wife Harriet. Fewkes, a Harvard-trained anthropologist who had been excavating the ruins at Awatobi, invited the Sykeses to attend the Niman Ceremony at First Mesa, which Emma described in her diaries. Later she also attended dances at the nearby Indian School ("I had some quite good waltzers") and made the acquaintance of the photographer Adam Clark Vroman.

The sense of a richly textured life comes through in Emma's journals, to which Godfrey occasionally adds a playful note. When she describes how the couple sets out on horseback in the cool of the evening, cutting across the high tableland through stands of juniper and piñons, Godfrey writes in the margin of her notebook that she will surely make a "jolly good rider," particularly if she continues to wear her "fetching cap."

It's difficult not to succumb to a romantic view of Emma Sykes's honeymoon. She is in love, lying in the hammock on the verandah of a charming house reading the latest periodicals. She has access to a library, servants, and the stunning beauty of an other-worldly landscape. She is privileged to observe the rich native culture. Still, one muses that it was Rachel Lee who had the benefit of greater time in which to absorb the deep experience of both native culture and pristine landscape, as she traversed the plateau on horseback over and over again, often alone, and spent those evenings with Tuba and his wife feasting on precious melons.

When Emma's stay at Keams Canyon ended, she and Godfrey made the eighty-mile trip to Flagstaff, their new home. By then Emma was pregnant, and at the age of forty-one she delivered her first child, Glenton Godfrey Sykes. Four years later she had another boy, Gilbert Walmisley. Her great-granddaughters, Susan Lowell and Diane Grua, recount that the marriage "was a happy one. . . but the rigors of such a life undoubtedly contributed to her early death, from rheumatic heart disease, in 1906 [after just over ten years in her adopted home in the American West]. Godfrey, Glenton, and Gilbert Sykes all lived well into their eighties, mainly in or near Tucson, where the family had relocated in 1906 for the sake of Emmie's health."

Emma Walmisley and Godfrey Sykes.
Courtesy of the Sykes Family.

Both Lucy Flake and Rachel Lee became lifelong residents of the plateau. Lucy preceded her husband in death and is buried in the cemetery in Snowflake. Rachel outlived her husband by thirty-five years. She spent most of those years in and around Safford, Arizona. She died in 1912 at the age of eighty-seven. Her last resting place is a little rock-strewn cemetery on a hill overlooking the desert she had come to know so well.

PEOPLE KEEP JOURNALS for many different reasons, and this is as true today as it was in the 19th century. Sometimes a diary is written with an eye to posterity, or to satisfy a longing to communicate the details of one's life. At other times, it's meant to be a place for private musings, a way of holding a solitary conversation with oneself.

It is interesting to note that Emma Sykes kept two diaries concurrently—one in the form of a long letter intended to be sent to relatives, the other for herself. In the former, she mentions the blue dress and the curling iron, perhaps as a way of reassuring her family that her comfort and needs were being seen to. In the latter, she does not mention these things. With all diaries, the question becomes, for whom was this written—others, or oneself? And most often the answer is both.

Still, one senses a great diversity of purpose in the diaries of Lucy Flake, Rachel Lee, and Emma Sykes. Lucy was recording not only the times in which she lived, but her feelings about polygamy, pioneer hardships, and family matters. Many early Mormons kept journals, in part because they believed they were engaged in the Great Gathering of Zion. In contrast, Rachel seems to have viewed herself as a humble keeper of the official record of her community. As for Emma, in her journal she became the rich chronicler of the New World, painting vivid word pictures for the benefit of the family she left behind in England.

More than a century has elapsed since these diaries were written. The journals of Lucy and Rachel and Emma are more than just records of individual lives. They are vivid and moving portraits of three remarkable women, and they add to the history of the region as well. In the end, what they say to us is this: listen to our voices, for we can tell you things.

After all, we were here where you are now.

JUDITH FREEMAN, novelist and critic, was born and raised in Ogden, Utah. She is the author of a collection of stories and three novels, including the acclaimed *Chinchilla Farm,* and is the great-granddaughter of Prudence Flake. A recipient of a Guggenheim Fellowship, she is presently on an extended stay in Rome, Italy, while she works on a novel of the Southwest.

ULRIKE ARNOLD:

Earth Painter

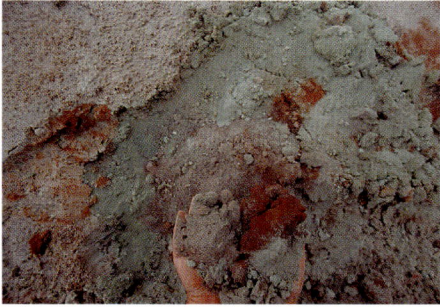

What fascinates me is how the earth, in the various forms of its landscape and stones, assumes different material and color qualities, how the variegated layers tell stories about the creation of the earth, about the conditions under which people live in different parts of the earth, as well as the interplay between climatic zones and vegetation.

The quality of the earth determines the quality of the landscape. It is a contributory factor for the architecture that people construct in the respective landscapes. In addition, the earth provides the basis for the production of colors, with which man develops his image of the world. In mysticism, too and among all peoples, the earth is assigned a particular significance: earth is primary materials, conveyor of energy; earth means beginning and end. — ULRIKE ARNOLD

In remote places on the Colorado Plateau, Ulrike Arnold, a German artist of international renown, creates works on canvas that convey a journalistic record and a sense of place, using not words but the materials at hand: the earth itself, the stuff of which the landscape around her is made. She creates her works on site, out of doors, and they capture with remarkable power a great deal about the place that surrounds her as she works.

Working on a piece of canvas which may be twenty feet long and six feet wide, Ulrike lays down colors that are not pigments but bits of soil, earth, sand, mud, and crushed rock. "I try to get the essence of a place," she says, and there is much detail in her painting. Her works have to do with color, but Ulrike also captures the structure and texture of the land where she paints. "It's all a matter of scale, you know; there is little in the big, and big in the little." She is inspired by the structure of the rocks, or the surface of a tree, all of the patterns and textures of nature which are, in themselves, abstract images to most of us.

FACING PAGE: Installation of earth flags, with earth paint. Flagstaff, Arizona, San Francisco Wash. 1993.

TOP: Earth paint on canvas, Bisbee, Arizona, 1998. Photo by Dave Edwards.

FACING PAGE: Earth paint on natural rock face, San Francisco Wash, 1993. Photo by Anselm Spring.

Ulrike Arnold's approach is an emotional rather than an intellectual one. And while her works effectively convey a sense of place, they are impressionistic and personal. Like written journals, they reveal as much about the person who creates them as they do about the place in which they are created.

Ulrike calls herself an "earth artist" or "earth painter," apt descriptions of the style in which she has worked for nearly twenty years. Born in Dusseldorf, she came to her craft first as a teacher of music and art. Most of her early creative work was done in pencil. She was drawn to the shadows of gray, the lines, the absence of artificial color. Then in 1980 she traveled to Provence, in France, where she visited the red ochre pits near Roussillion. It was a turning point in her life. She returned to her studio in Wuppertal with a small bit of that red earth.

Today Ulrike works entirely without a studio, and without artificial pigments. She has worked throughout the world, on every continent, often traveling alone, drawn to landscapes of startling beauty. She has worked in Algeria, Madagascar, Iceland, Armenia, and Australia, in the Canary Islands, and throughout the American Southwest. As she explores a new place, she asks, "What happened here thousands of years ago? Who lived here?" Her connection is not only with landscape but with the cultural heritage of the land, or as she calls it, "the spiritual meaning of landscape." Ulrike puts it most eloquently when she says, "Here must have been something special."

The most exciting place Ulrike ever worked was Ruby Gorge, in central Australia. In that remote place on her birthday in 1987, camping alone for the first time in her life, she transferred her feelings of fear, anxiety, joy, and excitement to the canvas. For her

"The theme of my pictures is the earth."

ABOVE: Luna Mesa, near Hanksville,
Utah. Photo by Anselm Spring.

FACING PAGE: Luna Mesa, near
Hanksville, Utah. Photo by Anselm Spring.

No one who knows Ulrike or her work can doubt that joy is the predominant emotion.

works, like most creative works of music or literature, are an outlet for loneliness, pain, suffering, and joy. No one who knows Ulrike or her work can doubt that joy is the predominant emotion.

Like the landscape itself, Ulrike's works are untitled and unframed, bearing only the names of the geographic locations where they were created and from which their earth paints derived. They are on display in galleries and museums throughout the world, the heavy, textured canvases loosely suspended from walls or spread horizontally on gallery floors.

In June 1999 Ulrike will exhibit at the Ludwigforum in Aachem, Germany. The show focuses on the relationship between nature and culture and will feature earth paintings of some twenty artists. For this venue Ulrike is preparing seven pieces, each twenty feet long and five feet wide. Each painting represents a separate location in the American Southwest, including San Francisco Wash, near Flagstaff; Bisbee, Arizona; the Burr Trail in Utah; Luna Mesa, near Capitol Reef National Park, Utah; Galisteo, New Mexico; and Guadalupe Ranch, in southern New Mexico.

In North America, Ulrike's work will be on display in August 1999 at the Nora Eccles Harrison Museum of Art in Logan, Utah. There her works will be shown along with those of Mario Reis, a German artist who works with water.

Ulrike feels strongly that her works are more than a mode of expression for the artist, more than a journal of her own feelings. They are done as a kind of mission, to inspire the viewer with feelings similar to those which she herself feels in places of overwhelming beauty. Viewers of her work will sometimes exclaim, "Can these colors be real, do they really exist in nature?" And Ulrike is delighted with the written comments she

ABOVE: Ulrike holding a large canvas at Notom, Utah. Photo by Anselm Spring.

FACING PAGE: Ulrike's most recent work, completed near Flagstaff in summer 1998. Photo by Dave Edwards.

receives along these lines: "You have opened my eyes to see the earth differently." Like the journals of Dave Edwards, Ulrike's works are meant for others to read.

Since 1991 Ulrike has experimented with painting directly on rock surfaces, first at the Crestone Zen Monastery in Colorado, later at the Christ in the Desert Monastery in Abiquiu, New Mexico, always on private land and only by invitation. Inspired by prehistoric rock art she has seen throughout the world, Ulrike feels her own earth works are places of "homage to creation, to nature." These works, sometimes carefully hidden in alcoves or niches, are always executed with a sensitivity to both the land and the people. She studiously avoids public land and sacred places. Like all her works, these works are created using only the natural materials she finds on site.

Ulrike hopes that her works invite a new vision of both the ordinary and the extraordinary, that they foster a sense of caring and stewardship toward the places they celebrate. They are her response to those places throughout the world where the landscape is so

remarkable, so startling, that words fail most of us in our attempts at description. Language will not serve us in our attempt to convey the power and glory of these places. The Colorado Plateau is one such place, and Ulrike Arnold has proven herself equal to that task, creating works that are an inspiration to all of us lucky enough to know them.

Tokens from the

THAR SHE BLOWS! A ceramic salt and pepper shaker commemorates the volcanic eruption of Mount St. Helens by "blowing its top" when a diner uses it to season a meal. The glaze is made with Mt. St. Helens ash. From Toutle, Washington.

BY THE SPOONFUL. Sterling silver state spoons of the late 19th century show lavish workmanship and materials. Commemorative spoons, plates, dishes, glasses, toothpick holders, and ashtrays provided take-home souvenirs that could be gathered state by state.

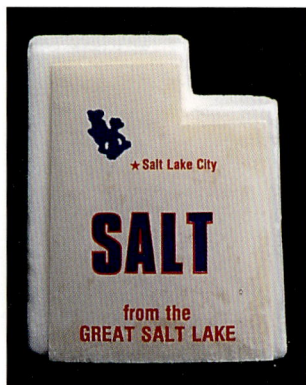

Road

Beautiful or bizarre, useful or not, artful or just plain tacky, tokens of place have fascinated footloose Americans from pre-automobile days to the present. From hand-beaded pincushions sold by Tuscarora Indians to Victorian tourists at Niagara Falls to the ubiquitous refrigerator magnets mass-produced in China and purveyed at every soft drink-and-gasoline stop along the interstate today, little reminders of a place and a moment have been picked up and packed along.

Like the jottings in a private notebook or a snapshot of the kids in front of a national monument, a salt shaker, a commemorative spoon, even a bamboo back scratcher, can become a talisman of one particular day and hour, a whole set of sights and smells, our fleeting presence at a crossroads of time and space that we will never know again. Even if we were to return, the place would not be the same — indeed, you truly can't step twice in the same river — and neither would we. We would at the very least be older — and maybe the cherished bit of bric-a-brac would no longer be for sale. So like jackdaws and bower birds we haul the booty of the moment back to our familiar nests where it can gather dust on our shelves, and maybe on the shelves of our descendants or of some perspicacious garage sale-goer of the 21st century. ~

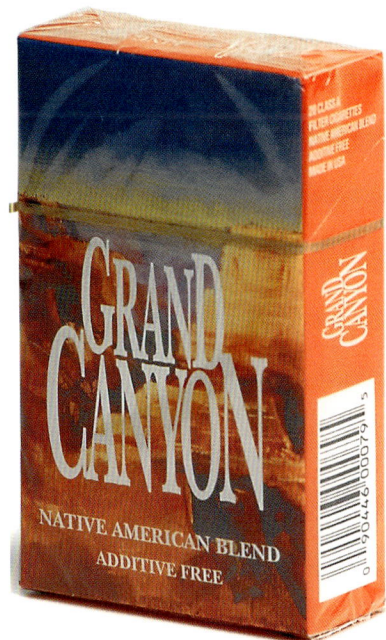

Cigarette packet, Harvey Girl figurine, Grand Canyon black velvet painting, and Grand Canyon tray shown courtesy of Grand Canyon National Park Museum Collection; all other objects from private collections, shown by permission.

DUCK ON THE ROCK

KAIBAB SUSPENSION BRIDGE

GRAND CANYON

BRIGHTY

ARIZONA

MULE TRAIN

KAIBAB TRAIL

GRAND CANYON NAT'L PARK

NAVAJO

TRAIL TO PHANTOM RANCH

NAVAJO WEAVER

INDIAN WATCHTOWER

BEEN THERE. . . National Park memorabilia built a visible record of family travels.

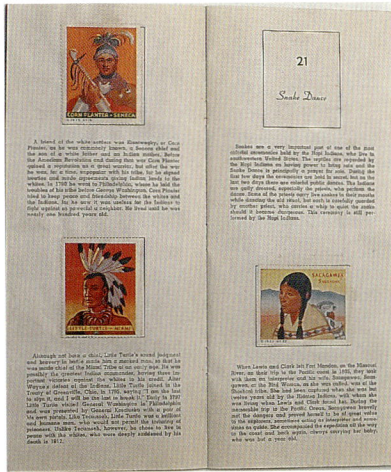

A TIME OF INNOCENCE? Boys and girls of the early 1930s and 1940s collected stamps from Diamond D-X gasoline service stations to fill their educational (and promotional) booklets on the topic of American Indians. "Skookums" Indian dolls, below, were collected in the Southwest. Collectibles like these create a record of changing attitudes and patterns of interaction between mainstream Americans and native tribal and other minority groups over the years.

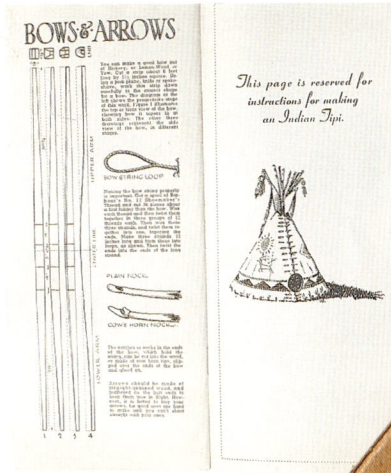

HAVING A GEM OF A TIME proclaims this postcard, which cost seven cents to mail in 1957.

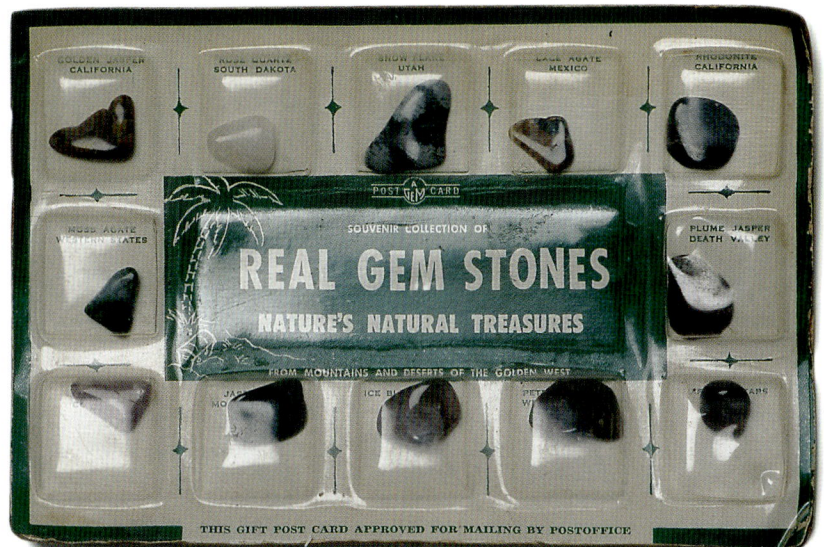

CHILDHOOD TREASURES.

Once purchased to placate rowdy, road-weary children, these trinkets, saved for decades, evoke cherished memories of family vacations.

ARCHES NATIONAL PARK

P.O. Box 907, Moab, UT 84532
Tel. 435-259-8161 (voice)
Tel. 435-259-5279 (TTY)

Arches National Park contains the world's largest concentration of natural stone openings. Red sandstone cliffs, spires, windows, and walls delight the eye and challenge the imagination. The park offers a great family experience. Paved roads guide visitors to spectacular views, and short easy hikes lead to fanciful formations.

ARIZONA STRIP INTERPRETIVE ASSOCIATION

345 East Riverside Drive, St. George,
UT 84790   Tel. 435-688-3275

The Arizona Strip Interpretive Association (ASIA) supports the Arizona Strip, a vast land stretching from the Nevada border to the Colorado River and from the Utah border to the Grand Canyon, boasting 4,000 miles of unpaved roads. Sales merchandise and Brown Bag Lectures are featured at the Bureau of Land Management (BLM) Interagency Information Center in St. George, Utah. The center is the place to buy maps, posters, and books concerning the Arizona Strip, Pine Valley Forest, Dixie BLM, Colorado Plateau, state and federal parks, and all recreational activities. Annual membership is $10.

BRYCE CANYON NATURAL HISTORY ASSOCIATION

Bryce Canyon National Park
Bryce Canyon, UT 84717
Tel. 435-834-5322

The mission of Bryce Canyon Natural History Association is to assist and promote the historical, scientific, and educational activities of Bryce Canyon National Park. It also supports the research, interpretation, and conservation programs of the National Park Service.

CANYONLANDS NATIONAL PARK

2282 South West Resource Boulevard,
Moab, UT 84532   Tel. 435-259-7164

Canyonlands National Park preserves an immense wilderness of rock at the heart of the Colorado Plateau. The Green and Colorado Rivers define the essence of this park. Canyonlands remains largely unspoiled today; its roads are mostly unpaved, its trails primitive, its rivers free-flowing.

CANYONLANDS NATURAL HISTORY ASSOCIATION

3031 South Highway 191, Moab, UT
84532   Tel. 435-259-6003

Canyonlands Natural History Association is a not-for-profit cooperating association operating in southeast Utah. CNHA is a partner to the land management agencies there in the development and marketing of high-quality interpretive materials which help to educate the visiting public as to sustainable land ethics.

CAPITOL REEF NATIONAL PARK

HC 70, Box 15, Torrey, UT 84775-9602
Tel. 435-425-3791

Capitol Reef National Park encompasses the Waterpocket Fold, a huge monocline eroded into a spectacular jumble of cliffs, domes, spires, canyons, and arches. The Fremont River cuts through the fold, supporting a diversity of plants and wildlife. Fremont Indian petroglyphs, historic structures and fruit orchards planted by Mormon settlers are preserved in the Fruita Historic District.

CAPITOL REEF NATURAL HISTORY ASSOCIATION

HC 70, Box 15, Torrey, UT 84775
Tel. 435-425-3791

Capitol Reef Natural History Association promotes the historical, cultural, scientific, interpretive, and educational activities and research of Capitol Reef National Park through the donation of proceeds from the sale of interpretive materials. Sales items are available at the Visitor Center and the newly renovated and refurnished Gifford farmhouse. The farmhouse, where visitors can purchase replicas of the past made by local artisans, serves as a cultural demonstration site to interpret the early Mormon settlement of the Fruita Valley.

COLORADO NATIONAL MONUMENT ASSOCIATION

c/o Colorado National Monument, Fruita,
CO 81521   Tel. 970-858-3617

Since 1964, Colorado National Monument Association (CNMA) has enhanced the visitor experience at Colorado National Monument, providing a marketplace for educational materials and vital financial support for research, interpretation, outreach, and staffing. CNMA publishes information about the northeast edge of the Colorado Plateau. CNMA members enjoy benefits that include a bimonthly newsletter, discounts, and invitations to special events.

DINOSAUR NATURE ASSOCIATION

1291 East Highway 40, Vernal,
UT 84078-2830   Tel. 435-789-8807

The Dinosaur Nature Association (DNA) is a not-for-profit organization created to aid the interpretive, educational, and scientific activities of the National Park Service at Dinosaur and Fossil Butte National Monuments. For more information or a catalog call 1-800-845-DINO (3466) or write to the above address.

DIXIE INTERPRETIVE ASSOCIATION

1696 Tamarisk Drive, Santa Clara, UT
84765   Tel. 435-628-3969

The Dixie Interpretive Association provides interpretive materials to increase awareness about multiple use management on the Dixie National Forest. Eighty percent of revenues go toward interpretive/educational programs and projects. Memberships are available. Association sales outlets are located in district offices in Cedar City, Panguitch, Escalante, and Teasdale, Utah. Visitor centers are located at Red Canyon (on Highway 12), Duck Creek (on Highway 14), and Wildcat (on Highway 12 south of Torrey).

DIXIE, MANTI-LASAL AND FISHLAKE NATIONAL FORESTS

82 North 100 East, Cedar City, UT 84720
Tel. 435-865-3700

The national forests of southern Utah— the Dixie, Manti-LaSal, and Fishlake—are "lands of many uses" that provide clean water, healthy fish and wildlife habitat, forage, timber, and outdoor recreation opportunities, from hiking and fishing to mountain biking, skiing, hunting, and water sports.

ENTRADA INSTITUTE / FRIENDS OF CAPITOL REEF

P.O. Box 750217
Torrey, UT 84775
Tel. 435-425-3265

The Entrada Institute / Friends of Capitol Reef celebrates the human and natural history of the Capitol Reef environs through the arts, humanities, and sciences. The institute sponsors a variety of courses for students of all ages and encourages a program of lifelong learning for seniors. Members receive discounts on classes, mailings, and a subscription to a semiannual newsletter.

GLEN CANYON NATURAL HISTORY ASSOCIATION

P.O. Box 581, Page, AZ 86040
Tel. 520-645-3532

We are dedicated people working in partnership to promote stewardship and inspire public awareness of resource issues affecting the Colorado Plateau. We join together for the support of educational, historical, and research projects within Glen Canyon and surrounding federal lands.

GRAND CANYON NATIONAL PARK

P.O. Box 129, Grand Canyon, AZ 86023
Tel. 520-638-7888

The Grand Canyon, a World Heritage Site, is recognized as a place of universal value, containing superlative natural and cultural features that are being preserved as part of the heritage of all the people of the world.

HUBBELL TRADING POST NATIONAL HISTORIC SITE

P.O. Box 150, Ganado, AZ 86505
Tel. 520-755-3475

Hubbell Trading Post National Historic Site is the oldest continuously operating trading post in the Navajo Nation and is the best remaining example of a traditional Southwest trading post. A crossroads of culture, the site preserves the past through its historic structures, Native American arts and objects, the landscape, and customs of another time.

KAIBAB NATIONAL FOREST

800 6th Street, Williams, AZ 86046
Tel. 520-635-8200

The Kaibab National Forest is an integral part of the heart of the Colorado Plateau. It surrounds the Grand Canyon and the communities of Williams, Parks, Tusayan, and Fredonia. One current focus is building relationships with the people interested in stewardship of these lands.

MESA VERDE MUSEUM ASSOCIATION

P.O. Box 38, Mesa Verde , CO 81330
Tel. 970-529-4445

The Mesa Verde Museum Association, a not-for-profit organization authorized by Congress and established in 1930, assists and supports the various interpretive programs, research activities, and visitor services of Mesa Verde National Park and Hovenweep National Monument. MVMA offers a membership program and supports the National Parks Electronic Bookstore at www.npeb.org

MESA VERDE NATIONAL PARK

P.O. Box 8, Mesa Verde National Park, CO
81330   Tel. 970-529-4475

Mesa Verde National Park, established in 1906, is one of the world's premier archaeological areas because of its high concentration of mesa-top structures and spectacular cliff dwellings. The park preserves the cultural heritage of the Ancestral Puebloan people, who lived in the area from A.D. 550 to A.D. 1300.

MUSEUM OF WESTERN COLORADO

P.O. Box 20,000-5020, Grand Junction,
CO 81502   Tel. 970-242-0971

The Museum of Western Colorado is open for public learning at four localities around Grand Junction: the Regional History Museum; the Research Center/Library; Dinosaur Valley; and Cross Orchards Historic Farm. The museum collects and preserves objects, specimens, and information about the fascinating cultural and natural history of western Colorado.

NORTHERN ARIZONA UNIVERSITY, CLINE LIBRARY

Box 6022, Flagstaff, AZ 86011-6022
Tel. 520-523-5551
www.nau.edu/library/speccoll/

As part of an educational institution, the Cline Library serves the Northern Arizona University academic community and the public by providing support for curricular, information, and research needs. The library's Special Collections and Archives Department collects, preserves, and makes available archival material which documents the history and development of the Colorado Plateau from prehistory to the present. The rich holdings represent a variety of disciplines and formats.

PEAKS, PLATEAUS AND CANYONS ASSOCIATION

c/o Tracey Hobson at Mesa Verde Museum Association, P.O. Box 38
Mesa Verde, CO 81330 Tel. 970-529-4445

Linked by both terrain and mission, the Peaks, Plateaus and Canyons Association (PPCA) is a group of not-for-profit educational associations, museums, and federal/state land management/tourism agencies. PPCA members seek to heighten visitor appreciation and understanding of the varied resources of the area on and around the Colorado Plateau. PPCA exchanges information and underwrites projects that advance the collective mission.

PETRIFIED FOREST NATIONAL PARK

P.O. Box 2217, Petrified Forest, AZ 86028-2217 Tel. 520-524-6228

Petrified Forest National Park is a respite in time where fossils of Triassic creatures and petrified wood converge with archaeological sites and historic settlements. This arid plateau also reveals the diversity of color and texture which are trademarks of the Painted Desert landscape. The park is a fascinating place where an ancient past meets the modern world.

PETRIFIED FOREST MUSEUM ASSOCIATION

P.O. Box 2217, Petrified Forest, AZ 86028-2217 Tel. 520-524-6228

Petrified Forest Museum Association is a not-for-profit organization established in 1914 to assist the interpretive, resource management, and educational programs at Petrified Forest National Park. Proceeds from the sale of publications are used to provide free information handouts, support scientific research, environmental education, Jr. Ranger programs, and special events.

PUBLIC LANDS INTERPRETIVE ASSOCIATION

6501 Fourth Street NW, Suite I, Albuquerque, NM 87107
Tel. 505-345-9498

Public Lands Interpretive Association (PLIA) provides visitor information services for the Bureau of Land Management, U.S. Forest Service, and the U.S. Fish & Wildlife Service throughout the Southwest. PLIA's Public Lands Information Centers offer comprehensive information about all public lands at a centralized location.

USDI BUREAU OF LAND MANAGEMENT

Utah State Office, P.O. Box 45155, Salt Lake City, UT 84145 Tel. 801-539-4223

The United States Department of the Interior's Utah Bureau of Land Management (BLM) is a federal land management agency responsible for managing, protecting, and improving 22 million acres of public lands in the state. Resources contained on these public lands include recreation, range, timber, minerals, watershed, fish and wildlife, wilderness, air, scenic, scientific, and cultural values. BLM is committed to caring for these resources in such a manner as to serve the needs of the American people for all times.

WUPATKI, SUNSET CRATER VOLCANO, AND WALNUT CANYON NATIONAL MONUMENTS

6400 North Highway 89
Flagstaff, AZ 86004 Tel. 520-526-1157

Wupatki, Sunset Crater Volcano, and Walnut Canyon National Monuments are managed collectively by the National Park Service. These units preserve and interpret outstanding examples of natural and cultural resources, elements of which help us to understand the larger story of the Colorado Plateau.

ZION NATURAL HISTORY ASSOCIATION

Zion National Park
Springdale, UT 84767 Tel. 435-772-3264

Zion Natural History Association (ZNHA) is a not-for-profit corporation working in cooperation with the National Park Service. The association funds interpretive projects, scientific research, and free publications for park visitors through sales of publications, maps, and other interpretive items and members' support. ZNHA members receive a discount on sales items and a semi-annual newsletter.

FEATURED MAP: EARLY SETTLEMENTS ON THE PLATEAU

any of the places mentioned in the journals of early settlers on the Colorado Plateau are located on the map at right. A map can illustrate the area defined as the Colorado Plateau, and can pinpoint the places where community has been created in the midst of its broad distances, but it cannot depict the experiences of adventurers and pioneers, creating new lives in strange new places or living fully in the familiar present. For that, we are fortunate to have the journals, diaries, and notebooks contained in this issue of the *Journal*. And in a sense these are also maps — word maps of human experience.

INSIDE BACK COVER: U.S. digital topographic map provided courtesy of Chalk Butte Inc. of Boulder, Wyoming. Customization by Kim Buchheit.